Synthesis Lectures on Engineering, Science, and Technology

The focus of this series is general topics, and applications about, and for, engineers and scientists on a wide array of applications, methods and advances. Most titles cover subjects such as professional development, education, and study skills, as well as basic introductory undergraduate material and other topics appropriate for a broader and less technical audience.

Mohammad Javad Karimi ·
Catherine Dehollain · Alexandre Schmid

Integrated Wireless Power, Data Communication, and Thermal Sensing Systems for Autonomous Multisite Brain Implants

Mohammad Javad Karimi
Swiss Federal Institute of Technology
Lausanne, Switzerland

Catherine Dehollain
Swiss Federal Institute of Technology
Lausanne, Switzerland

Alexandre Schmid
Swiss Federal Institute of Technology
Lausanne, Switzerland

ISSN 2690-0300   ISSN 2690-0327 (electronic)
Synthesis Lectures on Engineering, Science, and Technology
ISBN 978-3-031-90838-5   ISBN 978-3-031-90839-2 (eBook)
https://doi.org/10.1007/978-3-031-90839-2

© The Editor(s) (if applicable) and The Author(s), under exclusive license to Springer Nature Switzerland AG 2026

This work is subject to copyright. All rights are solely and exclusively licensed by the Publisher, whether the whole or part of the material is concerned, specifically the rights of translation, reprinting, reuse of illustrations, recitation, broadcasting, reproduction on microfilms or in any other physical way, and transmission or information storage and retrieval, electronic adaptation, computer software, or by similar or dissimilar methodology now known or hereafter developed.
The use of general descriptive names, registered names, trademarks, service marks, etc. in this publication does not imply, even in the absence of a specific statement, that such names are exempt from the relevant protective laws and regulations and therefore free for general use.
The publisher, the authors and the editors are safe to assume that the advice and information in this book are believed to be true and accurate at the date of publication. Neither the publisher nor the authors or the editors give a warranty, expressed or implied, with respect to the material contained herein or for any errors or omissions that may have been made. The publisher remains neutral with regard to jurisdictional claims in published maps and institutional affiliations.

This Springer imprint is published by the registered company Springer Nature Switzerland AG
The registered company address is: Gewerbestrasse 11, 6330 Cham, Switzerland

If disposing of this product, please recycle the paper.

To the beating hearts that shaped my dreams, my beloved parents…

Preface

Implanted medical devices (IMDs) have been widely developed to support the monitoring and recording of biological data inside the body or brain. Wirelessly powered IMDs, a subset of implantable electronics, have been proposed to eliminate the limitations related to the physical size constraints, battery replacement issues, and the complexity of wired powering. Wireless power transmission (WPT) methods have been employed to transfer wireless power and data between the implanted units and an external medical unit. Ultrasonic, capacitive, optical, radio frequency, and inductive links are the most commonly used WPT methods, according to the literature. Inductive power transfer is widely developed in wirelessly powered IMDs due to its robustness, design simplicity, and safety. Implantable neural interfaces are an example of wirelessly powered IMDs to monitor and detect neurological disorders. To achieve this, an autonomous wirelessly powered system is required that has the capacity of recording neural signals, transferring them to the base station wirelessly, detecting seizures in real time, and taking a responsive action or carrying out electrical stimulation to suppress the seizure from its initial onset. Consequently, an efficient and low-power wireless system is essential with the capability of multisite remote powering, bi-directional data communication, power management and control, and temperature sensing. This book presents a novel system-level concept by employing inductive links and CMOS electronics: a fully implantable wirelessly powered closed-loop multisite implant.

The proposed system has a three-layer architecture to support autonomous and closed-loop operations: (1) external unit, (2) central implanted unit (CIU), and (3) autonomous smart patches (ASPs). An external unit, embedded into a headstage, includes a system controller, battery, telemetry, and inductive link for power and bi-directional data transmission to the implant. When the battery is empty and power must be transmitted to the implant or data must be dumped from the memory of the implant, the headstage must be temporarily worn by the patient. The CIU housed in a burr hole receives the wireless power, to recharge the battery, and downlink data from the headstage. The CIU further supplies wireless power and configuration data to ASPs. The base station (CIU) is also responsible for controlling the entire implanted system, and managing power and temperature and data of the unit as well as the ASPs. The third layer of the proposed architecture consists of two ASPs that are implanted on the cortex surface. This book includes the analysis of the system-level modeling of the entire system using MATLAB Simulink, wireless power and data communication circuits in a CMOS technology, on-chip power control and temperature sensing, and system-on-chip measurement results to verify the proposed system.

A hybrid powering solution is considered for the CIU, consisting of wireless power transmission and a rechargeable battery. The required power of the ASPs is only supplied from the CIU using inductive links. A dual-band inductive link is employed at the CIU to provide multisite WPT to the two ASPs at two different frequencies. A power conversion chain (PCC) is designed in the CIU and ASPs to provide a stable DC voltage supply from the received AC signal. The PCC includes a full-wave active rectifier, a voltage reference, and low dropout (LDO) regulators. Moreover, a control unit and an automatic resonance tuning system are designed in the system to monitor and optimize remote powering, respectively. Since the CIU serves as the primary unit for WPT to the ASPs, a wireless power transfer unit is presented. Bi-directional data communication between the external unit and the CIU and between the CIU and ASPs is required. Consequently, low-power data communication units are proposed, using different modulation schemes. The external unit sends the frequency shift keying (FSK) data to the CIU and receives load shift keying (LSK) data from the CIU. Also, the ASPs receive amplitude shift keying (ASK) data from the CIU and send LSK data back to the CIU. Furthermore, an on-chip CMOS temperature sensor is presented to provide on-chip temperature sensing and monitor temperature elevation inside the brain implants. The entire system including both CIU and ASPs is integrated and fabricated using 180 nm standard CMOS technology and the performance of the full system is measured and verified. This work was supported by the Swiss National Science Foundation (SNSF) under Grant 200020-182548.

Lausanne, Switzerland
February 2025

Mohammad Javad Karimi
Catherine Dehollain
Alexandre Schmid

Competing Interests The authors have no competing interests to declare that are relevant to the content of this manuscript.

Contents

1 Introduction 1
   1.1 Study of Wirelessly Powered Biomedical Implants 1
      1.1.1 Commercial Biomedical Devices 3
      1.1.2 Safety Concerns 3
   1.2 Wireless Powering via Inductive Links 4
      1.2.1 Single Inductive Link 5
      1.2.2 Multiple Inductive Links 6
   1.3 Data Communication Methods in Biomedical Implants 7
      1.3.1 Modulation Schemes 7
      1.3.2 Downlink Data Communication 9
      1.3.3 Uplink Data Communication 9
   1.4 Discussion 11
   1.5 General Aim of the Book 11
      1.5.1 System Overview 14
   1.6 Contributions of This Book 17
   1.7 Outline of the Book 18
   References 19

2 System-Level Modeling 25
   2.1 System Overview 25
   2.2 Wireless Power System-Level Modeling 27
      2.2.1 Full-Wave Active Rectifier 27
      2.2.2 Low-Dropout Voltage Regulator (LDO) 27
      2.2.3 Power Feedback and Power-on-Reset (PoR) 29
   2.3 Data Communication System-Level Modeling 29
      2.3.1 Downlink Communication Model 30
      2.3.2 Uplink Data Communication Model 31

2.4	Recording and Stimulation System Architecture	37
	2.4.1 Signal Pre-processing in the Analog Domain	37
	2.4.2 Biomedical Digital Signal Processing	37
	2.4.3 Neural Stimulator and Charge Balancer	38
2.5	Summary	39
References		40

3 Implantable Dual-Band Inductive Link 43
   3.1 General Context 43
   3.2 Dual-Band Coil Designing Method 44
   3.3 Measurement Results 47
      3.3.1 Parallel Configuration 49
      3.3.2 Overlapping Configuration 50
      3.3.3 Frequency Response and PTE 50
   3.4 Summary 51
   References 52

4 Wireless Power Conversion Chain and Control Methods 55
   4.1 System Overview 55
   4.2 Full-Wave Active Rectifier 57
      4.2.1 Bias Circuit 58
      4.2.2 Offset-Controlled Common-Gate Comparator 59
      4.2.3 Logic Unit 60
      4.2.4 Measurement Results 60
   4.3 Voltage Reference 61
      4.3.1 Circuit Design 62
      4.3.2 Measurement Results 63
   4.4 Low-Dropout (LDO) Voltage Regulator 63
      4.4.1 1.8 V LDO with Off-Chip Output Capacitor 64
      4.4.2 1.2 V LDO with Capacitorless Output 66
      4.4.3 Measurement Results 67
   4.5 Power Control Unit 68
      4.5.1 Power Feedback (PF) 69
      4.5.2 Power on Reset (PoR) 71
      4.5.3 Voltage Limiter (VL) 72
      4.5.4 Measurement Results 72
   4.6 Wireless Power Transfer Unit 75
      4.6.1 Power Amplifier (PA) 76
      4.6.2 Relaxation Oscillator 77
      4.6.3 DC-DC Converter 78
      4.6.4 Measurement Results 79

	4.7	Automatic Resonance Tuning System	80
		4.7.1 Properties of Inductive Links	83
		4.7.2 Capacitor Bank	86
		4.7.3 Start-Up and Termination	86
		4.7.4 Clock Recovery and Divider	87
		4.7.5 Control Unit	88
		4.7.6 Measurement Results	91
		4.7.7 Discussion	94
	4.8	Summary	95
	References		97
5	**Wireless Data Communication**		103
	5.1	Clock Recovery (CR)	104
	5.2	LSK Modulator	105
	5.3	Separated-Vb ASK Demodulator	105
		5.3.1 Circuit Architecture	105
		5.3.2 Measurement Results	111
	5.4	Averaging ASK Demodulator	113
		5.4.1 Circuit Architecture	113
		5.4.2 Measurement Results	113
	5.5	FSK Demodulator	115
		5.5.1 Circuit Architecture	115
		5.5.2 Measurement Results	118
	5.6	Summary	119
	References		120
6	**Towards On-Chip CMOS Temperature Sensing**		123
	6.1	CMOS Sensor Architecture	124
		6.1.1 Bandgap Voltage Reference	128
	6.2	10-Bit SAR ADC Design	129
	6.3	Sensor Simulation Results	130
	6.4	Measurement Results	130
	6.5	Temperature Sensor for Body Temperature Range	134
	6.6	Summary	135
	References		136
7	**System-On-Chip Integration**		139
	7.1	Background and Related Research	139
	7.2	System Architecture	142
	7.3	Measurement Results	144

	7.4 Discussion	148
	7.5 Summary	152
	References	153
8	**Summary and Conclusions**	155

Acronyms

ADC	Analog to Digital Converter
AFE	Analog Front End
ART	Automatic Resonance Tuning
ASK	Amplitude Shift Keying
ASP	Autonomous Smart Patch
BER	Bit Error Rate
BGR	Bandgap Reference
BMI	Brain Machine Interface
CGC	Common-Gate Comparator
CIU	Central Implanted Unit
CMP	Comparator
CTAT	Complementary to Absolute Temperature
DAC	Digital to Analog Converter
DCC	Data Conversion Chain
EA	Error Amplifier
ESD	Electrostatic Discharge
FSK	Frequency Shift Keying
GBW	Gain Bandwidth
HDP	High Delivered Power
IMD	Implanted Medical Device
IPT	Inductive Power Transfer
ISM	Industrial, Scientific, and Medical
LDO	Low Dropout Regulator
LDP	Low Delivered Power
LDR	Load Regulation
LHP	Left Half Plane
LNR	Line Regulation
LSK	Load Shift Keying

MCU	Microcontroller
MEA	Microelectrode Array
OOK	On-Off Shift Keying
OTA	Operational Transconductance Amplifier
PA	Power Amplifier
PCB	Printed Circuit Board
PCC	Power Conversion Chain
PCE	Power Conversion Efficiency
PDL	Power Delivered to the Load
PF	Power Feedback
PoR	Power-on-Reset
PSK	Phase Shift Keying
PSR	Power Supply Rejection
PSRR	Power Supply Rejection Ratio
PTAT	Proportional to Absolute Temperature
PTE	Power Transfer Efficiency
RF	Radio Frequency
R-OSC	Relaxation Oscillator
SAR	Successive Approximation Register
SoC	System-on-Chip
UWB	Ultra Wideband
VCR	Voltage Conversion Ratio
VL	Voltage Limiter
VNA	Vector Network Analyzer
WPDT	Wireless Power and Data Transmission
WPT	Wireless Power Transmission

Introduction

1

Implanted medical devices (IMDs) are developed to monitor and record biological data from the units implanted inside the body or brain and send it to an external unit. The need for IMDs has considerably increased in recent years [1, 2] to be used for diagnostic purposes. The applications of these implantable biomedical devices include cardioverter defibrillators, cochlear implants, and deep brain stimulators. Wirelessly powered implants are a subset of IMDs that employ wireless power transmission techniques to avoid using batteries or wired powering. This method enables the design of compact and long-lasting IMDs for biomedical applications. In neural applications, IMDs are used for neural recording and brain stimulation. For instance, brain implants are utilized for Parkinson's disease and epilepsy, by providing targeted open- or closed-loop stimulation to certain areas of the brain [2–4]. Ultrasonic, capacitive, optical, radio frequency (RF), and inductive links are the most commonly used platforms employed as a wireless power transmission technique. Inductive power transfer (IPT) is widely developed in IMDs because it is robust, straightforward, and safe. It also enables simultaneous wireless power and data communication. This book presents an implantable system that employs inductive links and CMOS electronics to develop a fully implantable wirelessly powered closed-loop multisite implant.

1.1 Study of Wirelessly Powered Biomedical Implants

Wireless power transmission (WPT) methods have been extensively explored to enable IMDs for biomedical applications, as depicted in Fig. 1.1. These methods facilitate the transfer of power from an external source through air or material to devices implanted in the body, encompassing both near-field and far-field techniques. By utilizing WPT, IMDs reduce

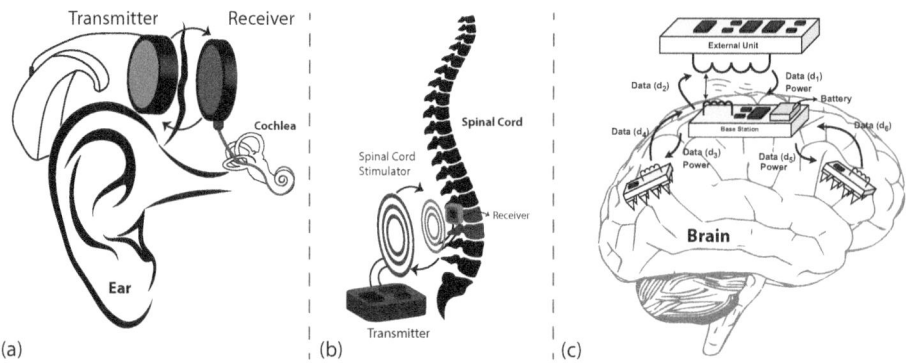

Fig. 1.1 Conceptual illustration of an external power unit paired with implantable medical devices: **a** cochlear implant, **b** spinal cord stimulator, and **c** brain implant

the risk of infection commonly associated with traditional wired power sources. Recent technological advancements, including the miniaturization of semiconductor components, have significantly accelerated progress in this area [3]. IMDs serve a wide range of purposes, such as monitoring physiological parameters and enhancing the functionality of specific body parts. Over the past few decades, they have become fundamental in managing health conditions and tracking patients' biological data [4]. To fulfill these roles, IMDs require a reliable and continuous power supply with high power transfer efficiency (PTE) and sufficient power delivery to the load (PDL) [5]. Power sources for IMDs can be broadly categorized into two types [3]: (1) internal (in-vivo) sources, which operate independently of external input, such as batteries (rechargeable or single-use), supercapacitors, or energy harvested from the human body; and (2) external (ex-vivo) sources, which provide power remotely via WPT technologies.

Size and power consumption are critical constraints for devices implanted within the body and brain. Internal power sources, such as batteries, occupy substantial space and present challenges related to maintenance—requiring recharging for rechargeable batteries or replacement for non-rechargeable ones. Furthermore, temperature variations in energy consumption can lead to issues such as overheating, chemical leakage, or, in extreme cases, explosions. Given the stringent size limitations of the human body and its sensitivity to elevated temperatures, external power sources and wireless power transmission methods are often preferred, despite their relatively lower power efficiency. With increasing frequencies, the tissue absorption and power loss increase, and thus more heat is produced, which implies a hazard. For higher frequencies, the wavelength reduces and thus we may enter the far field. Hence, short-range power and data transmission methods are widely used for IMDs due to their reduced tissue absorption, minimized power loss, and improved safety compared to far-field techniques. This chapter focuses on short-range wireless power and data transmission facilitated by an external source. Communication between the external unit and IMDs can be categorized into three primary functionalities, as shown in Fig. 1.2a. Systems may implement

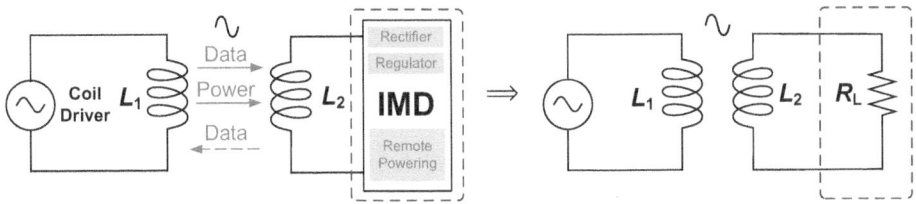

Fig. 1.2 Simplified block diagram of a wireless power transmission (WPT) system

all or a subset of these functionalities: (1) remote power transmission from the external unit to the IMDs, (2) transmission of data or commands from the external unit to the internal unit to activate the IMD, configure its settings, or stimulate specific body or brain regions (downlink or forward telemetry), and (3) feedback transmission from the internal unit to the external unit to monitor system-related parameters (e.g., received power level, temperature, and electrode impedance) and recorded biological data (uplink or back telemetry).

1.1.1 Commercial Biomedical Devices

Wirelessly powered brain and body implants are a rapidly growing field of research and development in the industry too. However, their practical influence has so far been somewhat restricted. This limitation can be attributed in part to the complexity of the human brain, obstacles in validating devices, concerns regarding their long-term reliability, and the absence of highly effective technologies essential for broad adoption [6]. In recent years, there has been a significant increase in the number of commercial devices both in key industry players (e.g., Medtronic, Boston Scientific, and Abbott) and neurotech startups that incorporate wirelessly powered implants or wearable devices, such as those used for deep brain stimulation, cochlear implants, and pacemakers. The closed-loop responsive neurostimulation (RNS) system provides four channels of neural recording for epilepsy detection [7]. Percept PC from Medtronic utilizes spectral analysis for brain stimulation [8]. The brain-computer interface (BCI) designed by Synchron enables patients with limited hand mobility to control devices with their thoughts. BCIs designed by Paradromics and Blackrock Neurotech [9, 10] enable multi-electrode recording and bidirectional data communication. The brain implant designed by Neuralink can record neural signals and provide stimulation.

1.1.2 Safety Concerns

Safety considerations pose significant challenges for wireless power and data transmission in biomedical applications. Various standards have been established to regulate WPT through the human body, differing by country and subject to periodic updates. First, materials used in implantable devices must be biocompatible to avoid tissue rejection [3]. Additionally, the effects of electromagnetic fields on the body and brain must be evaluated based on factors such as frequency, tissue sensitivity, absorption rates, and power density. For example, absorbed waves can induce heating in the surrounding tissues, cause radiation effects, and potentially stimulate muscles and nerves. Tissue absorption leading to temperature increases is proportional to the frequency and input power, necessitating limitations on both to minimize heating effects. When exposed to an electromagnetic field, the energy absorbed per unit mass of tissue is quantified as the specific absorption rate (SAR), measured in watts per kilogram (W/kg). International standards, such as those from IEEE, recommend that SAR values should not exceed 2 W/kg for 10 g of tissue [11]. In the literature, SAR values for inductive power transfer systems are reported at 0.4 W/kg for an input power of 15 mW [5] and 0.25 W/kg for 5.55 mW [12].

Another critical safety concern is the heating caused by power dissipation and losses, which can be harmful when tissue temperatures exceed normal body temperature by several degrees for extended periods. The permissible temperature increase is generally limited to 1–2 °C [13]. The allowable dissipated power varies by tissue type and must be assessed individually. For instance, a study on temperature rise and allowable power densities for various tissues is available in [14]. Factors such as coil losses (e.g., the Eddy effect), loose coupling due to coil misalignment or distance, and mismatches in electronic circuits contribute to power losses, temperature elevation, and reduced efficiency. Maintaining appropriate PTE and PDL values is essential to control these effects. The literature suggests a maximum allowable power dissipation of 40 mW/cm^2 in the body to ensure temperature increases remain below 2 °C [13].

1.2 Wireless Powering via Inductive Links

Several methods have been developed for wireless power transmission, the most notable techniques being: (1) inductive link, (2) ultrasound link, (3) optical link, (4) capacitive link, and (5) electromagnetic link. Each method offers specific advantages and limitations based on the intended application. Ultrasonic links transmit power by converting electrical signals into ultrasound waves, which are received by an implanted piezoelectric sensor [3]. This approach achieves high-efficiency power transfer with deep tissue penetration, though it is unsuitable for transmission across the skull due to significant attenuation through bone [15]. Additionally, beamforming is necessary when powering multiple implants wirelessly. Optical links use light emission for power transfer, offering high data rates and immunity to electromagnetic interference. However, their limitations include low penetration depth, sensitivity to misalignment, and high absorption in biological tissues [5]. Capacitive links utilize two parallel plates to transfer power over very short distances, typically a few millimeters, with a maximum distance of up to 10 mm as reported in [16]. These links, along with inductive links, are advantageous for compact implantable systems as they require min-

1.2 Wireless Powering via Inductive Links

imal area. Among the various methods, inductive and ultrasound links exhibit higher power transfer efficiency compared to other techniques. From a safety perspective, inductive links demonstrate lower tissue wave absorption at higher input power levels relative to other WPT methods. This section explores inductive power transmission techniques, including single and multiple inductive links. A literature review highlights that the fundamental concept of an IPT system involves a pair of coils positioned at a short distance. As previously mentioned, the magnetic flux generated by the primary coil induces an alternating voltage in the secondary coil, which provides the necessary power for the implanted devices. However, IPT systems face challenges with low power efficiency due to weak inductive coupling. To address this, capacitors are added to both the primary and secondary coils to enhance power transfer efficiency, compensate for the secondary leakage inductance (a key factor in poor coupling condition) [17], and ensure resonance at the carrier frequency.

1.2.1 Single Inductive Link

Using a single pair of coils is the most straightforward approach to designing a system capable of simultaneously transmitting wireless power and data, as illustrated in Fig. 1.3. Various WPT systems employing a single pair of coils have been discussed in the literature [18–20]. This method minimizes the interfaces between links, thereby reducing exposure to electromagnetic fields and making it safer and more reliable for the human body compared to other WPT techniques. However, achieving simultaneous power and data transmission presents challenges due to the trade-off required between power efficiency and data rate [3]. To address this, specific modulation techniques and methodologies must be employed. A single inductive link, while straightforward, has limitations such as weak coupling between coils and sensitivity to misalignment, resulting in low efficiency and non-uniform power distribution [21]. Power efficiency is largely influenced by the quality factor (Q) of the inductances [22], whereas the data rate is determined by the system's bandwidth (BW). Therefore, a key design consideration is the relationship between the bandwidth and quality factor, expressed as $BW = f_c/Q$, where BW is the bandwidth, Q is the quality factor, and f_c is the center frequency. This relationship is valid when the inductor and capacitor resonate at the specified resonance frequency.

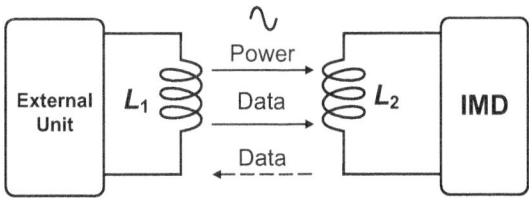

Fig. 1.3 Topology of an IMD with a single inductive link

1.2.2 Multiple Inductive Links

Multiple links, as illustrated in Fig. 1.4, represent an alternative technique for simultaneously transferring power and data, offering enhanced system performance and higher data rates compared to single-inductive links. In this approach, separate coils are designated for each function. Data transmission is divided into two categories: downlink and uplink. Downlink transmission is used for sending commands, such as initiating a stimulation phase, while uplink transmission sends biological parameters or reports the performance of the device, such as voltage or power levels, to an external unit. In systems employing multiple inductive links, one link is typically dedicated to power transfer and another to data communication, whether for downlink or uplink. Some designs use a single link for bidirectional data communication [23–26], while others allocate separate links for each direction of data communication [27]. In certain implementations, antennas and RF waves are utilized for back telemetry, enabling data transmission from the implanted device to the external unit, as depicted in Fig. 1.4b. While this approach provides higher data rates and improved robustness, it also consumes more power and introduces greater design complexity compared to single-inductive link systems. Additionally, minimizing cross-talk between the coils poses a significant challenge.

Table 1.1 summarizes the advantages and drawbacks of each WPT technique, categorized into two main groups: single and multiple inductive links. Each group includes two possible topologies. Additionally, the table highlights methodologies proposed in the literature to address or mitigate the limitations of each technique, along with their corresponding references. For example, in the case of single inductive links, alternative modulation schemes have been proposed to overcome low data rates and the challenges associated with simultaneous power and data transmission. For multiple links, various coil geometry structures are suggested to reduce cross-talk by employing different shapes. Table 1.2 presents a compar-

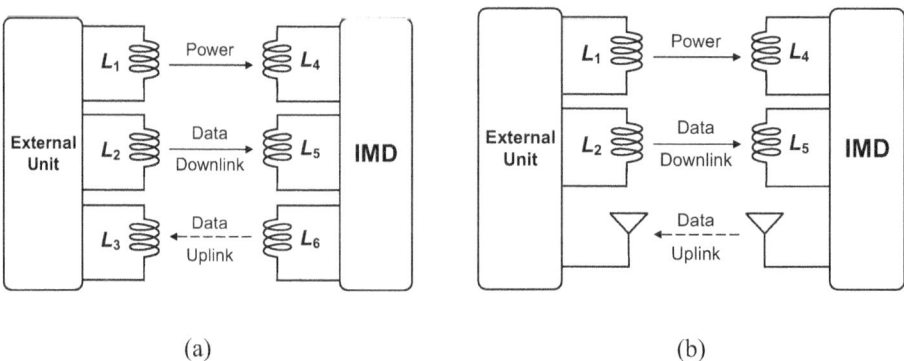

Fig. 1.4 Topologies involving multiple inductive links: **a** three separated inductive links and **b** RF/inductive link

1.3 Data Communication Methods in Biomedical Implants

Table 1.1 Summarized comparison of advantages and drawbacks of inductive powering methods [2]

Methods		Advantages	Drawbacks	Improvement [Refs.]
Single link	– Single-pair of coils	– Saving space – Safety – Low-power – High-PTE – Simple structure	– Power discontinuity – Challenges in power and data sending simultaneously	– CWM[a] [28] – QCWM[b] [29] – COOK[c] [30] – PPSK[d] [19]
	– With additional resonators			
Multiple links	– Separate links	– Continuous power – High data rate – Longer distance	– Cross-talk – High-power – Safety issue – Complex structure	– Coplanar coils [27] – Figure-8 coils [31] – Orthogonal coils [32]
	– RF/Inductive			

[a] Carrier Width Modulations (CWM)
[b] Quadrature CWM (QCWM)
[c] Cyclic On-Off Keying (COOK)
[d] Passive PSK (PPSK)

ison of WPT systems based on key parameters, including carrier frequency, the number of additional resonators, coil distances, PTE, and PDL.

1.3 Data Communication Methods in Biomedical Implants

The primary goal of power transmission to IMDs is to provide the energy required for sending and receiving data to or from the body. This section introduces modulation schemes, detailing their respective advantages and drawbacks, and categorizes prior studies based on these techniques.

1.3.1 Modulation Schemes

Modulation schemes are critical for data communication as they enable reliable data transfer while protecting against noise and disturbances [41, 42]. The most common schemes include amplitude shift keying (ASK), phase shift keying (PSK), and frequency shift keying (FSK), each with its variations. In ASK, the signal amplitude is modulated, offering simple circuitry and low power consumption. However, it is less resilient to external factors like noise, coupling variation, and interference, which can affect carrier amplitude and limit data rates. The modulation index (MI) in ASK is defined as $MI = (V_H - V_L)/(V_H + V_L)$, where V_H and

Table 1.2 Performance and characteristics of WPT systems [2]

References	Year	IPT link	n_{res}^a	Detail	f_{power}^b (MHz)	d^c (mm)	PTE (%)	PDL (mW)
[5]	2019	Single	1–2	3 or 4-coil	13.56	5	22.70	122
[33]	2018	Multiple	–	2-coil + RF (data)	8	10	30.00	10
[34]	2017	Single	1	3-coil	60	16	2.40	1.3
[12]	2017	Multiple	1	3 separate links	131	18	2.01	100
[19]	2016	Single	–	Single-pair of coils	13.56	8	60.64	100
[35]	2016	Single	–	Single-pair of coils	13.56	70	5.80	5.4
[36]	2016	Single	4	6-coil	13.56	70	14.00	42
[37]	2015	Multiple	1	3-coil + RF (data)	13.56	10	41.60	25
[20]	2015	Single	–	Single-pair of coils	13.56	30	50.00	102
[38]	2015	Single	2	4-coil	13.56	40	69.00	120
[39]	2014	Multiple	1–2	3 or 4-coil + RF	13.56	120	12.60	20
[40]	2010	Single	–	Single-pair of coils	13.56	10	6.90	11.2

[a] Number of additional resonators between Tx-Rx
[b] Carrier frequency for WPT
[c] Tx-Rx distance

V_L represent the highest and lowest voltage amplitudes, respectively. A special case of ASK, On-Off Keying (OOK), involves $MI = 100$ ($V_L = 0$) to enhance immunity against disturbances [43]. However, OOK cannot provide continuous power during '0' bits. In contrast, PSK modulates the signal phase, while FSK uses two carrier frequencies. Both schemes offer higher immunity to noise, higher data rates, and continuous power transmission but at the cost of higher power consumption and greater design complexity compared to ASK [29].

1.3.2 Downlink Data Communication

In stimulation applications, downlink communication allows the external unit to send commands to determine which body or brain regions should be stimulated or measured. According to [4], downlink communication can be categorized into: (1) single-carrier and (2) multi-carrier data telemetry. In the single-carrier approach, the same carrier is used for both data modulation and power transmission, simplifying circuit and coil design, enhancing reliability, and maintaining compactness of the device. Modulation schemes like ASK, FSK, and PSK are commonly used in this scenario. To address data rate limitations of single-carrier systems, multi-carrier techniques are employed, using separate carriers for downlink and uplink data communication. This allows higher PTE and data rates by dedicating a specific coil or antenna to each function (e.g., power, downlink, and uplink), operating at distinct frequencies. However, this method introduces complexity in design, link interference, increased bandwidth requirements, and higher power consumption.

1.3.3 Uplink Data Communication

Uplink data communication is essential for transmitting system parameters (e.g., delivered power, voltage, and temperature) and biological values (e.g., impedance) to the external unit. It is divided into two categories: (1) passive and (2) active data telemetry [4, 44]. In passive telemetry, data is transmitted back using the same magnetic or electromagnetic field that supplies power. This reduces power consumption as the carrier is generated externally and shared for both wireless power transfer and data communication. When electromagnetic coupling is used (far-field), this technique is referred to as backscattering, while in magnetic coupling (near-field), it is called load modulation. In load modulation, a switch connected to the secondary coil adjusts the load, altering the voltage or current in the primary coil and enabling data recovery. For active telemetry, modulation schemes such as ASK, PSK, FSK, and wideband techniques like impulse radio ultra-wideband (IR-UWB) are commonly employed. These methods are preferred for achieving higher data rates in uplink communication, often utilizing compact antennas.

Table 1.3 Comparative analysis of the wireless data communication systems [2]

References	Tech(μm)	Link[a]	Method[b]	Power f_{power} (MHz)	Downlink f_{down} (MHz)	Downlink Data (Mbps)	Downlink Modulation	Uplink f_{up} (MHz)	Uplink Data (Mbps)	Uplink Modulation	P_c[c] (mW)	BER	Area (mm^2)
[45]	0.5	M	2	2	4/8	2	FSK	70–200	2	OOK	14.4	–	2.3
[41]	0.18	M	2	2	2	1	ASK	0.2–0.33	0.133	FSK	0.481	–	–
[25]	0.18	M	2	2	20	2	DPSK	20	2	LSK	18	2×10^{-7}	29.5
[46]	0.5	M	2	2.64	2.64	0.0065	ASK	433	0.33	FSK	13.5	–	27.73
[23]	0.8	M	2	5	5	1.25	FSK	50–100	3	ASK/BPSK	–	–	2.32
[33]	0.18	M	3	8	8	0.5	ASK	416	12	OOK	0.544	–	–
[47]	0.5	M	2	13.56	13.56	1.69	BPSK	0.1	0.1	LSK	5	–	0.1
[48]	0.35	M	2	13.56	50	13.56	PDM[d]	50	13.56	PDM[d]	–	4.3×10^{-7}	1.6
[49]	0.18	M	3	13.56	13.56	0.0066	PPM[e]	869	1.5	OOK	0.65	–	0.161
[37]	0.18	M	3	13.56	2400	100	OOK	3100–7000	500	UWB	10.4	10^{-5}	0.8
[19]	0.6	S	1	13.56	13.56	0.4	OOK	13.56	1.35	PPSK	–	5.98×10^{-8}	–

[a] Multiple (M) or Single (S) inductive link
[b] Communication method in downlink/uplink: (1) single inductive, (2) inductive/inductive, (3) inductive/RF
[c] Power consumption of the implanted circuit
[d] Pulse delay modulation
[e] Pulse harmonic modulation

1.4 Discussion

Table 1.3 presents a comparative review of systems incorporating both downlink and uplink data communication. The comparison includes key parameters such as the type of modulation, CMOS technology, power link frequency, carrier frequency, data rate, power consumption of the implanted CMOS circuit, bit error rate (BER), and the area of the implanted CMOS circuit. Table 1.4 offers a comprehensive review of recent wireless power and data transmission systems utilizing inductive links for biomedical applications. Based on the table, the maximum PTEs are observed within systems featuring a distance between the transmitter and receiver in the range of 10–20 mm, with a power carrier frequency equal to or less than 13.56 MHz. Additionally, the inclusion of supplementary coil resonators improves power transfer efficiency. The PTE experiences a decline with an increase in carrier frequency or Tx-Rx distance, given that the dimensions of the coils remain constant.

As per the collected data rates, carrier-less pulse-based and impulse-radio ultra-wideband (IR-UWB) modulation schemes are suggested for achieving data rates surpassing 10 Mbps in both uplink and downlink data communication, at the expense of higher power consumption. Data rates, on average, are higher within frequency ranges from 10 to 100 MHz compared to lower frequency ranges. Furthermore, systems using multiple links exhibit higher data rates than systems that rely on a single inductive link, due to the link's trade-off between the quality factor and bandwidth in the inductive link to achieve higher PTE and data rate. However, this advantage increases system complexity and potential interference issues.

1.5 General Aim of the Book

This book aims at presenting modern and original methods to develop a multisite implantable wireless system for neural recording and stimulation inside the brain. A novel layered implantable system is presented that supports all presented original developments. The proposed system consists of wireless power and data units, providing continuous and simultaneous power and data transmission along with power control and temperature sensing units. Inductive powering is used as a WPT technique. Figure 1.5 shows the conceptual view of the wirelessly powered system inside the brain.

The proposed system consists of a three-layer architecture, deemed the appropriate method to provide patients with autonomy. An external base unit is embedded into a headstage, and includes a system controller, battery, telemetry and antennas for power and bidirectional data transmission to the implant. The headstage is assumed to be only temporarily worn by the patient, in times when power must be transmitted to the implant or data must be dumped from the implant. The internal base station housed in a Burr hole receives power and data from the headstage, and further wirelessly delivers them to autonomous smart patches. In addition, the internal base station controls the entire implant, and manages the power/temperature/data of the unit as well as the smart patches. The third layer consists of

Table 1.4 Comprehensive assessment of wireless power and data transmission characteristics for IMDs [2]

References	Year	Tech (μm)	Link	Power f_{power} (MHz)	Downlink f_{down} (MHz)	Downlink Data (Mbps)	Downlink Modulation	Uplink f_{up} (MHz)	Uplink Data (Mbps)	Uplink Modulation	d (mm)	PTE (%)	P_c (mW)	BER	PDL (mW)	a (mm^2)
[50]	2019	–	Multiple	433	–	–	–	0.5	0.125	CDMA	12.5	–	1.05	–	–	–
[51]	2019	0.18	Single	144	144	0.015	ASK	–	–	–	10	3.4	2.7	–	–	9
[5]	2019	0.35	Single	13.56/60	60	–	OOK	60	–	LSK	5	22.70	0.544	–	122	1.1
[33]	2018	0.18	Multiple	8	8	0.5	ASK	416	12	OOK	10	30	18	2.00E–07	10	–
[25]	2017	0.18	Multiple	2	20	2	DPSK	20	2	LSK	–	–	0.111	5.00E–06	–	29.5
[12]	2017	0.35	Multiple	131	131	–	ASK	131	1	OOK	18	2.01	0.0145	–	100	0.2
[28]	2017	0.13	Single	27.12	27.12	9.04	CWM	–	–	–	–	–	6.51	1.00E–06	–	0.0018
[52]	2017	0.13	Multiple	1.5	1.5	–	ASK	3.1–10.6 G	46	UWB	20	–	0.0355	–	–	6
[29]	2016	0.13	Single	27.12	27.12	10.85	QCWM	–	–	–	–	–	–	–	–	–
[30]	2016	0.065	Single	13.56	–	–	–	13.56	6.78	COOK	10	–	–	9.90E–08	11.5	0.92
[53]	2016	0.18	Single	10	–	–	–	10	2	LSK	9	–	–	4.79E–04	15	–
[19]	2016	0.6	Single	13.56	13.56	0.4	OOK	13.56	1.35	PPSK	8	60.64	–	5.98E–08	100	–
[37]	2015	0.18	Multiple	13.56	2400	100	OOK/BPSK	3.1–7 G	500	UWB	10	41.60	10.4	1.00E–05	25	0.8
[54]	2015	0.35	Multiple	13.56	–	–	–	–	–	LSK	70	36.30	24	–	–	3.5
[35]	2015	0.35	Single	13.56	–	–	–	13.56	0.25	LSK	20	13.50	–	–	–	2.54
[20]	2015	0.35	Single	13.56	–	–	–	13.56	0.5	LSK	30	50.00	–	–	102	–

(continued)

1.5 General Aim of the Book

Table 1.4 (continued)

References	Year	Tech (μm)	Link	Power f_{power} (MHz)	Downlink f_{down} (MHz)	Downlink Data (Mbps)	Downlink Modulation	Uplink f_{up} (MHz)	Uplink Data (Mbps)	Uplink Modulation	d (mm)	PTE (%)	P_c (mW)	BER	PDL (mW)	a (mm²)
[49]	2014	0.18	Multiple	13.56	13.56	0.0066	PPM	869	1.5	OOK	30	16.80	0.65	–	–	0.161
[39]	2014	–	Multiple	13.56	–	–	–	–	–	LSK	120	12.60	–	–	–	–
[48]	2014	0.35	Multiple	13.56	50	13.56	PDM	50	13.56	PDM	10	–	–	4.30E−07	42	1.6
[55]	2013	0.35	Multiple	–	66.6	20	PHM	66.6	20	PHM	10	–	–	8.70E−08	–	0.23
[56]	2013	0.18	Single	8.4	–	–	–	8.4	1	ASK	15	36	–	–	10	–
[23]	2012	0.8	Multiple	5	5	1.25	FSK	50–100	3	ASK/BPSK	20	–	–	–	–	2.32
[41]	2011	0.18	Multiple	2	2	1	ASK	0.2–0.33	0.133	FSK	10	–	0.481	–	–	–
[57]	2011	0.5	Multiple	–	67.5	10.2	PHM	67.5	10.2	PHM	10	–	–	6.30E−08	–	–
[58]	2010	0.18	Multiple	2	22	2	DPSK	–	–	–	10	–	5.7	1.00E−04	100	27.03
[27]	2010	–	Multiple	1	13.56	4.16	OQPSK	13.56	4.16	OQPSK	–	61.00	–	2.00E−06	–	–
[59]	2010	0.5	Multiple	13.56	–	–	–	915/433	–	PWM/FSK	70	–	5.85	–	–	16.17
[60]	2009	0.5	Multiple	13.56	13.56	0.1	BPSK	13.56	–	LSK	–	–	2.3	–	–	0.28
[45]	2009	0.5	Multiple	4/8	4/8	2	FSK	70–200	2	OOK	10	–	14.4	–	–	2.3
[47]	2008	0.5	Multiple	13.56	13.56	1.69	BPSK	0.1	0.1	LSK	20	–	5	–	22.5	0.1
[61]	2007	–	Multiple	0.5	5/10	0.2	FSK	400	–	PWM	10	–	–	–	–	–
[46]	2007	0.5	Multiple	2.64	2.64	0.0065	ASK	433	0.33	FSK	130	13	13.5	–	–	27.73
[62]	2005	–	Multiple	0.125	25/50	–	FSK	2450	–	RF	–	–	–	–	–	–
[63]	2005	1.5	Multiple	1	–	–	–	1	0.003	LSK	7	65.80	250	1.00E−07	–	–
[64]	2005	0.18	Single	10	10	1.12	BPSK	10	–	LSK	15	–	0.7	1.00E−05	0.61	0.2

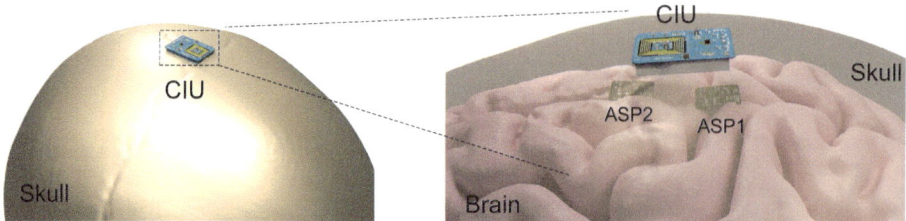

Fig. 1.5 Conceptual view of the remotely powered system

smart patches that are implanted on the cortex surface. Each smart patch records the electrical activity of a cortical zone through up to 16 electrodes and the amplification and signal conditioning electronics, detects seizures using digital feature extractors and electrically stimulates the cortex to suppress the seizure. The three-layer architecture is original in the research field and is the key to providing autonomy to cortical implantable systems.

The major research progresses highlighted in this book include methods and developments pertaining to:

- Modeling the full system, including power, thermal and data constraints, and developing the appropriate management strategies to guarantee the reliable operation of the three-layer system architecture.
- Development of implantable dual-band inductive link for wireless power and data transmission.
- Development of implanted telemetry system delivering multisite power to smart patches located on the surface of the cortex, and bidirectional data communication.
- Design of an on-chip integrated temperature sensor.
- Integration of the system with experimental results.

1.5.1 System Overview

Figure 1.6 shows a block diagram of the full system that is proposed, consisting of external and implanted systems. Implantable patches are shown in A, which consist of autonomous systems including a microelectrode array, a processing unit with the capability of recording, from the electrodes, feature extraction, stimulation channels, in addition to wireless power and data communication to a central control unit in B. The central control unit in B includes power and data communication to two implanted smart patches, a general control microcontroller (MCU) in charge of thermal and power management of the entire implanted system, memory and its data management, and a power conversion and management chain

1.5 General Aim of the Book

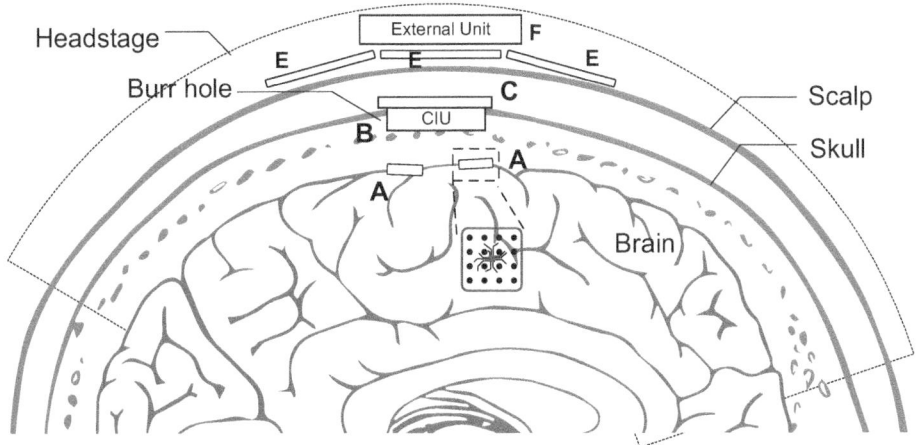

Fig. 1.6 Overview of the wirelessly powered system

also including an implanted battery. Part C is the telemetry system that receives power and communicates data from and to the external base station in F; C also includes the implanted antenna.

Modules B and C are implanted into the Burr hole, which is the prerequisite to any surgical craniotomy. Part C is implanted into a limited-size excavation of the surface of the skull that receives the antenna. All external parts are supported by a lightweight headstage that mechanically holds the external base station F (EBS) and antennas for power and data transmission to the implanted parts in E. The headstage is worn on a temporary basis that is patient-dependent. The necessity to wear the headstage is dictated by a low level power, or an excessive amount of data in the internal memory. Hence, the headstage must be worn to recharge the internal batteries, and to dump recorded data into an infrastructure memory. These extreme cases are signaled to the patient through an alarm signal that is delivered to the patient. In terms of transfer of power, the headstage is connected to power to recharge its batteries inside F. When it is worn by the patient, power is delivered to the implanted system in B which in turn recharges its implanted battery which is of lower capacity than the external battery. Eventually, a power management module in B and A dictates the necessity to transfer power from the central implanted unit (CIU) in B towards the autonomous smart patches (ASP) in A. Control data is wirelessly transferred from the external unit in F into the CIU in B, which in turn wirelessly transfers it to the ASP in A. Information data as well as cortical signals of importance to the clinician who wants to screen them, are wirelessly transferred from the ASP in A to the CIU in B. In the absence of the headstage, cortical data are stored in a compressed way in a local memory.

This fundamental working hypothesis is proposed as a way to significantly improve over existing and proposed systems as follows, and has the following specificities:

- Power and data are transferred in three stages, instead of the generally proposed two stages. Autonomy is a benefit of this specific architecture which allows storage of power and data inside the intermediate stage, while the external headstage can be removed by the patient.
- The implanted system is active at all times, resulting in no data loss, and no lack of power to record and stimulate.

The hypothesis establishes various constraints on the system and blocks, which ultimately lead to the specification of microelectronic and RF development.

- A multistage power management system must be developed that does not only consider local and instantaneous power dissipation, but integrates information from the diverse blocks into a power management strategy.
- The implanted system modules A, B, and C are autonomous, and must include a battery to store power, a power management system, as well as a data communication platform.
- Thermal dissipation must be considered.

The methods and developments presented in this book are limited to the implanted parts that operate wirelessly. Figure 1.7 presents the block-level description of the proposed system. The implanted part, which is the focus of the proposed project, comprises three major modules. The internal power management and data modules are in charge of all data and power transmission and conversion, i.e., to the external unit as well as to ASPs. Temperature sensors in the implanted patches deliver local temperature to the MCU.

An internal telemetry unit receives power from the external unit and transfers data in a bidirectional manner. A power conversion chain (PCC) is in charge of generating appropriate voltage for the digital, analog and stimulation power domains. The cortical signal is recorded from a multichannel system (14–16 channels), and is conditioned and converted into the digital domain by an analog front-end (AFE) unit. Eventually, epilepsy features are extracted

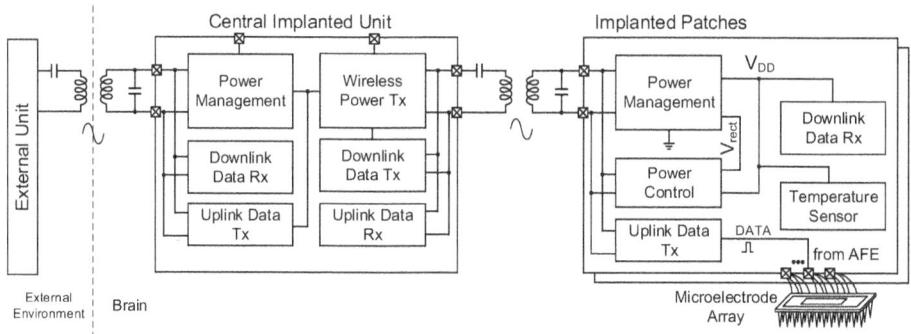

Fig. 1.7 Block diagram of the proposed system

using a digital block implementing a certified algorithm in the first phase. Stimulation is eventually carried out over stimulation channels. The designed temperature sensor also wirelessly sends its temperature to the internal base station in order to enable control of the power delivered by the internal station to the ASP. Moreover, the local controller can monitor the temperature to reduce the received power in the ASP. An ASP integrated circuit (IC) is developed such as to be integrated within one dedicated microelectrode array, which also embeds the telemetry coils.

Some issues related to the packaging and the capacity of implantable batteries are recognized but are considered outside the scope of this book. The significant development of implantable systems reasonably allows expecting that battery packaging for medical-grade systems will improve and be solved within the next years. Also, the immense need for autonomy enables expecting batteries of higher capacities as well as memory of larger capacities to be available in the future. These developments are expected at a horizon of five to ten years and will match the expected maturity of the proposed system. The system was designed and fabricated using a 180 nm standard CMOS technology due to its cost-effectiveness and ability to provide suitable core and IO voltages for power management and data communication circuits.

1.6 Contributions of This Book

The upcoming chapters will present the blocks of the entire system, as shown in Figs. 1.7 and 1.8. The main contribution of this book includes designing a multisite wireless power and telemetry system with a dual-band inductive link. The system includes a wireless power conversion unit, data communication unit, on-chip temperature sensor, and automatic resonance tuning system. The primary innovation of this project focuses on the system level, emphasizing the necessity of designing a full system. This goal is achieved by designing some circuits based on well-established architectures, implementing small enhancements and customizations tailored to the specifications and requirements of the project. Simultaneously, certain circuits and designed blocks are entirely original and novel, presenting improvements in power efficiency or transient performance. The entire system is validated with silicon fabrication and measurements.

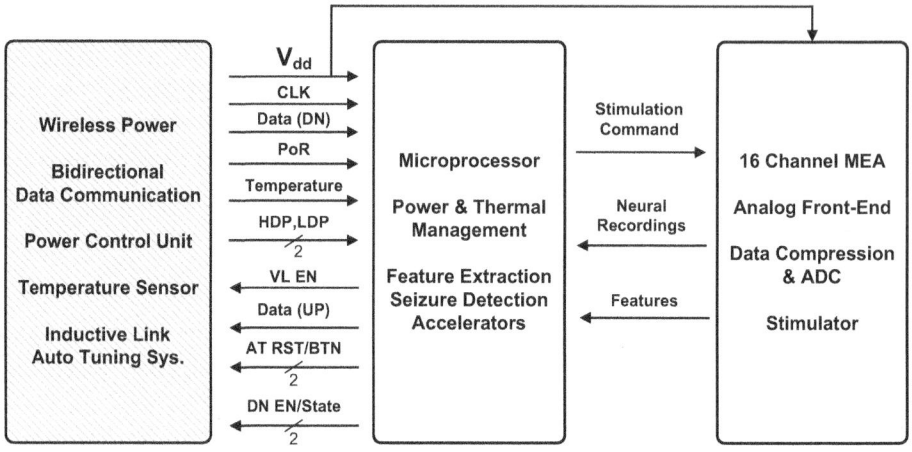

Fig. 1.8 Overview of different operating blocks of implanted part of the proposed system with the highlighted designed block as the main contributions of this book

1.7 Outline of the Book

This book is organized as follows, as shown in Fig. 1.9:

- In Chap. 2, a system-level modeling and analysis for a wireless power and data transmission system is carried out, including a closed-loop epileptic seizure detection for low-power implantable applications.
- In Chap. 3, a dual-band inductive link is proposed to operate at both 6.78 and 13.56 MHz in the industrial, scientific, and medical (ISM) bands.

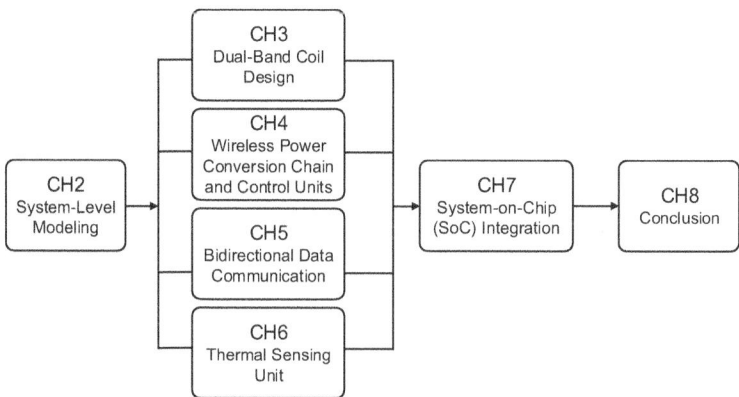

Fig. 1.9 Chapter organization in the book

- In Chap. 4, the wireless power supply, conversion mechanisms within implanted medical devices, and control units are designed.
- In Chap. 5, a low-power bi-directional data communication system and a clock generation are presented.
- In Chap. 6, a fully integrated CMOS temperature sensor is presented, which uses CTAT/PTAT voltage comparison with a sensitivity improvement circuit.
- In Chap. 7, the overall system's effectiveness is measured. The proposed circuits were integrated as a system-on-chip (SoC) and fabricated using a standard 180 nm MM/RF CMOS technology. The measurement results of the SoCs are also presented.
- In Chap. 8, summary and potential future research are discussed.

References

1. K. Kwon, J. U. Kim, S. M. Won, J. Zhao, R. Avila, H. Wang, K. S. Chun, H. Jang, K. H. Lee, J.-H. Kim, S. Yoo, Y. J. Kang, J. Kim, J. Lim, Y. Park, W. Lu, T. il Kim, A. Banks, Y. Huang, and J. A. Rogers, "A battery-less wireless implant for the continuous monitoring of vascular pressure, flow rate and temperature," *Nature Biomedical Engineering*, vol. 7, pp. 1215–1228, Apr. 2023.
2. M. J. Karimi, A. Schmid, and C. Dehollain, "Wireless Power and Data Transmission for Implanted Devices via Inductive Links: A Systematic Review," *IEEE Sensors Journal*, vol. 21, no. 6, pp. 7145–7161, 2021.
3. A. Trigui, S. Hached, A. C. Ammari, Y. Savaria, and M. Sawan, "Maximizing Data Transmission Rate for Implantable Devices over a Single Inductive Link: Methodological Review," *IEEE Reviews in Biomedical Engineering*, vol. 12, pp. 72–87, 2019.
4. B. Lee and M. Ghovanloo, "An Overview of Data Telemetry in Inductively Powered Implantable Biomedical Devices," *IEEE Communications Magazine*, vol. 57, pp. 74–80, 2019.
5. Y. Jia, S. A. Mirbozorgi, P. Zhang, O. T. Inan, W. Li, and M. Ghovanloo, "A Dual-Band Wireless Power Transmission System for Evaluating mm-Sized Implants," *IEEE Transactions on Biomedical Circuits and Systems*, vol. 13, pp. 595–607, 2019.
6. M. Shoaran, "Next-Generation Closed-Loop Neural Interfaces: Circuit and AI-driven innovations," *IEEE Solid-State Circuits Magazine*, vol. 15, no. 4, pp. 41–49, 2023.
7. "NeuroPace — neuropace.com." www.neuropace.com. [Accessed on 02-2025].
8. "Health tech for the digital age — medtronic.com." https://www.medtronic.com/. [Accessed on 02-2025].
9. "Enabling the future — paradromics.com." https://www.paradromics.com/. [Accessed on 02-2025].
10. "Blackrock Neurotech — blackrockneurotech.com." https://blackrockneurotech.com/. [Accessed on 02-2025].
11. Ieee, "IEEE Standard for Safety Levels with Respect to Human Exposure to Radio Frequency Electromagnetic Fields, 3 kHz to 300 GHz Amendment 1: Specifies Ceiling Limits for Induced and Contact Current, Clarifies Distinctions between Localized Exposure and Spatial Peak Power Density," *IEEE Std C95.1a-2010 (Amendment to IEEE Std C95.1-2005)*, pp. 1–9, 2010.
12. P. Yeon, S. A. Mirbozorgi, J. Lim, and M. Ghovanloo, "Feasibility Study on Active Back Telemetry and Power Transmission Through an Inductive Link for Millimeter-Sized Biomedical Implants," *IEEE Transactions on Biomedical Circuits and Systems*, vol. 11, pp. 1366–1376, 2017.

13. P. D. Wolf, "Thermal considerations for the design of an implanted cortical brain-machine interface (BMI)," *Indwelling Neural Implants: Strategies for Contending with the in Vivo Environment*, vol. 3, pp. 63–86, 2007.
14. T. Samaras, A. Christ, A. Klingenböck, and N. Kuster, "Worst case temperature rise in a one-dimensional tissue model exposed to radiofrequency radiation," *IEEE Transactions on Biomedical Engineering*, vol. 54, pp. 492–496, 2007.
15. M. Meng and M. Kiani, "Design and Optimization of Ultrasonic Wireless Power Transmission Links for Millimeter-Sized Biomedical Implants," *IEEE Transactions on Biomedical Circuits and Systems*, vol. 11, pp. 98–107, 2017.
16. A. Koruprolu, S. Nag, R. Erfani, and P. Mohseni, "Capacitive Wireless Power and Data Transfer for Implantable Medical Devices," *2018 IEEE Biomedical Circuits and Systems Conference, BioCAS 2018 - Proceedings*, vol. 2018, pp. 1–4, 2018.
17. K. Van Schuylenbergh and R. Puers, *Inductive Powering: Basic Theory and Application*, vol. 151. Dordrecht: Springer Netherlands, 2009.
18. M. J. Karimi, M. Jin, Y. Zhou, C. Dehollain, and A. Schmid, "Wirelessly Powered and Bi-directional Data Communication System with Adaptive Conversion Chain for Multisite Biomedical Implants Over Single Inductive Link," *IEEE Transactions on Biomedical Circuits and Systems*, pp. 1–11, 2024.
19. D. Jiang, D. Cirmirakis, M. Schormans, T. A. Perkins, N. Donaldson, and A. Demosthenous, "An Integrated Passive Phase-Shift Keying Modulator for Biomedical Implants with Power Telemetry over a Single Inductive Link," *IEEE Transactions on Biomedical Circuits and Systems*, vol. 11, pp. 64–77, 2017.
20. X. Li, C. Y. Tsui, and W. H. Ki, "A 13.56 MHz Wireless Power Transfer System With Reconfigurable Resonant Regulating Rectifier and Wireless Power Control for Implantable Medical Devices," *IEEE Journal of Solid-State Circuits*, vol. 50, pp. 978–989, 2015.
21. D. Ahn and M. Ghovanloo, "Optimal Design of Wireless Power Transmission Links for Millimeter-Sized Biomedical Implants," *IEEE Transactions on Biomedical Circuits and Systems*, vol. 10, pp. 125–137, 2016.
22. F. Asgarian and A. M., *Wireless Telemetry for Implantable Biomedical Microsystems*. IntechOpen: Laskovski, 2011.
23. A. D. Rush and P. R. Troyk, "A Power and Data Link for a Wireless-Implanted Neural Recording System," *IEEE Transactions on Biomedical Engineering*, vol. 59, no. 11, pp. 3255–3262, 2012.
24. M. J. Karimi, S. Mehdi, C. Dehollain, and A. Schmid, "Wireless Power and Data Transceiver in A Central Implanted Unit for Biomedical Applications," in *2024 IEEE 15th Latin America Symposium on Circuits and Systems (LASCAS)*, pp. 1–5, 2024.
25. Y. K. Lo, Y. C. Kuan, S. Culaclii, B. Kim, P. M. Wang, C. W. Chang, J. A. Massachi, M. Zhu, K. Chen, P. Gad, V. R. Edgerton, and W. Liu, "A Fully Integrated Wireless SoC for Motor Function Recovery after Spinal Cord Injury," *IEEE Transactions on Biomedical Circuits and Systems*, vol. 11, pp. 497–509, 2017.
26. M. J. Karimi, C. Dehollain, and A. Schmid, "Power Feedback Control Unit for Closed-Loop Wirelessly Powered Biomedical Implants," *IEEE Transactions on Circuits and Systems II: Express Briefs*, vol. 70, no. 5, pp. 1674–1678, 2023.
27. G. Simard, M. Sawan, and D. Massicotte, "High-speed OQPSK and efficient power transfer through inductive link for biomedical implants," *IEEE Transactions on Biomedical Circuits and Systems*, vol. 4, pp. 192–200, 2010.
28. A. Trigui, M. Ali, A. C. Ammari, Y. Savaria, and M. Sawan, "A 14.5 μw generic Carrier Width demodulator for telemetry-based Medical Devices," *Proceedings - 2017 IEEE 15th International New Circuits and Systems Conference, NEWCAS 2017*, vol. 2017, no. 15, pp. 369–372, 2017.

29. A. Trigui, M. Ali, A. C. Ammari, Y. Savaria, and M. Sawan, "Quad-Level Carrier Width Modulation demodulator for micro-implants," *14th IEEE International NEWCAS Conference, NEWCAS 2016*, vol. 2016, no. 14, pp. 1–4, 2016.
30. S. Ha, C. Kim, J. Park, S. Joshi, and G. Cauwenberghs, "Energy Recycling Telemetry IC with Simultaneous 11.5 mW Power and 6.78 Mb/s Backward Data Delivery over a Single 13.56 MHz Inductive Link," *IEEE Journal of Solid-State Circuits*, vol. 51, no. 11, pp. 2664–2678, 2016.
31. U. M. Jow and M. Ghovanloo, "Optimization of a multiband wireless link for neuroprosthetic implantable devices," *2008 IEEE-BIOCAS Biomedical Circuits and Systems Conference, BIOCAS 2008*, vol. 2008, pp. 97–100, 2008.
32. W. Liu, M. Sivaprakasam, G. Wang, M. Zhou, J. Granacki, J. Lacoss, and J. Wills, "Implantable biomimetic microelectronic systems design," *IEEE Engineering in Medicine and Biology Magazine*, vol. 24, no. 5, pp. 66–74, 2005.
33. R. Ranjandish, K. Ture, F. Maloberti, C. Dehollain, and A. Schmid, "All Wireless, 16-Channel Epilepsy Control System with Sub-μW/Channel and Closed-Loop Stimulation Using a Switched-Capacitor-Based Active Charge Balancing Method," *ESSCIRC 2018 - IEEE 44th European Solid State Circuits Conference*, vol. 2018, pp. 22–25, 2018.
34. S. A. Mirbozorgi, P. Yeon, and M. Ghovanloo, "Robust Wireless Power Transmission to mm-Sized Free-Floating Distributed Implants," *IEEE Transactions on Biomedical Circuits and Systems*, vol. 11, no. 3, pp. 692–702, 2017.
35. B. Lee, M. Kiani, and M. Ghovanloo, "A Triple-Loop Inductive Power Transmission System for Biomedical Applications," *IEEE Transactions on Biomedical Circuits and Systems*, vol. 10, pp. 138–148, 2016.
36. S. A. Mirbozorgi, Y. Jia, D. Canales, and M. Ghovanloo, "A Wirelessly-Powered Homecage with Segmented Copper Foils and Closed-Loop Power Control," *IEEE Transactions on Biomedical Circuits and Systems*, vol. 10, pp. 979–989, 2016.
37. S. A. Mirbozorgi, H. Bahrami, M. Sawan, L. A. Rusch, and B. Gosselin, "A Single-Chip Full-Duplex High Speed Transceiver for Multi-Site Stimulating and Recording Neural Implants," *IEEE Transactions on Biomedical Circuits and Systems*, vol. 10, pp. 643–653, 2016.
38. S. A. Mirbozorgi, H. Bahrami, M. Sawan, and B. Gosselin, "A Smart Cage With Uniform Wireless Power Distribution in 3D for Enabling Long-Term Experiments With Freely Moving Animals," *IEEE Transactions on Biomedical Circuits and Systems*, vol. 10, pp. 424–434, 2016.
39. U. M. Jow, P. McMenamin, M. Kiani, J. R. Manns, and M. Ghovanloo, "EnerCage: A smart experimental arena with scalable architecture for behavioral experiments," *IEEE Transactions on Biomedical Engineering*, vol. 61, pp. 139–148, 2014.
40. M. Kiani and M. Ghovanloo, "An RFID-based closed-loop wireless power transmission system for biomedical applications," *IEEE Transactions on Circuits and Systems II: Express Briefs*, vol. 57, pp. 260–264, 2010.
41. C. H. Kao, Y. P. Lin, and K. T. Tang, "Wireless data and power transmission circuits in biomedical implantable applications," *Proceedings of 2011 International Symposium on Bioelectronics and Bioinformatics, ISBB 2011*, pp. 9–12, 2011.
42. M. J. Karimi, Y. Zhou, C. Dehollain, and A. Schmid, "An Analysis of An ASK Demodulator With Dual Self-Biased Separated Voltages for Implantable Applications," *IEEE Transactions on Circuits and Systems II: Express Briefs*, pp. 1–1, 2024.
43. E. G. Kilinc, O. Atasoy, C. Dehollain, and F. Maloberti, "FoM to compare the effect of ASK based communications on remotely powered systems," in *2011 7th Conference on Ph.D. Research in Microelectronics and Electronics, PRIME 2011 - Conference Proceedings* (D. R. in Microelectronics and T. Electronics, eds.), pp. 29–32, 2011.
44. L. Yu, B. J. Kim, and E. Meng, "Chronically implanted pressure sensors: Challenges and state of the field," *Sensors (Switzerland)*, vol. 14, no. 11, pp. 20620–20644, 2014.

45. A. M. Sodagar, G. E. Perlin, Y. Yao, K. Najafi, and K. D. Wise, "An implantable 64-channel wireless microsystem for single-unit neural recording," *IEEE Journal of Solid-State Circuits*, vol. 44, pp. 2591–2604, 2009.
46. R. R. Harrison, P. T. Watkins, R. J. Kier, R. O. Lovejoy, D. J. Black, B. Greger, and F. Solzbacher, "A low-power integrated circuit for a wireless 100-electrode neural recording system," *IEEE Journal of Solid-State Circuits*, vol. 42, pp. 123–133, 2007.
47. S. Sonkusale and Z. Luo, "A complete data and power telemetry system utilizing BPSK and LSK signaling for biomedical implants," *Proceedings of the 30th Annual International Conference of the IEEE Engineering in Medicine and Biology Society, EMBS'08 - "Personalized Healthcare through Technology"*, vol. 2008, no. 30, pp. 3216–3219, 2008.
48. M. Kiani and M. Ghovanloo, "A 13.56-Mbps pulse delay modulation based transceiver for simultaneous near-field data and power transmission," *IEEE Transactions on Biomedical Circuits and Systems*, vol. 9, pp. 1–11, 2015.
49. E. G. Kilinc, C. Dehollain, and F. Maloberti, "A low-power PPM demodulator for remotely powered batteryless implantable devices," in *Midwest Symposium on Circuits and Systems*, (Tx), pp. 318–321, College Station, 2014.
50. P. Feng, M. Maslik, and T. G. Constandinou, "EM-Lens Enhanced Power Transfer and Multi-Node Data Transmission for Implantable Medical Devices," *BioCAS 2019 - Biomedical Circuits and Systems Conference, Proceedings*, vol. 2019, pp. 1–4, 2019.
51. C. Kim, J. Park, S. Ha, A. Akinin, R. Kubendran, P. P. Mercier, and G. Cauwenberghs, "A 3 mm*3 mm Fully Integrated Wireless Power Receiver and Neural Interface System-on-Chip," *IEEE Transactions on Biomedical Circuits and Systems*, vol. 13, pp. 1736–1746, 2019.
52. H. Kassiri, M. T. Salam, M. R. Pazhouhandeh, N. Soltani, J. L. Perez Velazquez, P. Carlen, and R. Genov, "Rail-to-rail-input dual-radio 64-channel closed-loop neurostimulator," *IEEE Journal of Solid-State Circuits*, vol. 52, pp. 2793–2810, Nov. 2017.
53. Y.-P. Lin, C.-Y. Yeh, P.-Y. Huang, Z.-Y. Wang, H.-H. Cheng, Y.-T. Li, C.-F. Chuang, P.-C. Huang, K.-T. Tang, H.-P. Ma, Y.-C. Chang, S.-R. Yeh, and H. Chen, "A Battery-Less, Implantable Neuro-Electronic Interface for Studying the Mechanisms of Deep Brain Stimulation in Rat Models," *IEEE Transactions on Biomedical Circuits and Systems*, vol. 10, no. 1, pp. 98–112, 2016.
54. B. Lee, M. Kiani, and M. Ghovanloo, "A Smart Wirelessly Powered Homecage for Long-Term High-Throughput Behavioral Experiments," *IEEE Sensors Journal*, vol. 15, pp. 4905–4916, 2015.
55. M. Kiani and M. Ghovanloo, "A 20-Mb/s pulse harmonic modulation transceiver for wideband near-field data transmission," *IEEE Transactions on Circuits and Systems II: Express Briefs*, vol. 60, pp. 382–386, 2013.
56. G. Yilmaz, O. Atasoy, and C. Dehollain, "Wireless energy and data transfer for in-vivo epileptic focus localization," *IEEE Sensors Journal*, vol. 13, pp. 4172–4179, 2013.
57. F. Inanlou, M. Kiani, and M. Ghovanloo, "A 10.2 Mbps pulse harmonic modulation based transceiver for implantable medical devices," *IEEE Journal of Solid-State Circuits*, vol. 46, pp. 1296–1306, 2011.
58. K. Chen, Z. Yang, L. Hoang, J. Weiland, M. Humayun, and W. Liu, "An integrated 256-channel epiretinal prosthesis," *IEEE Journal of Solid-State Circuits*, vol. 45, pp. 1946–1956, 2010.
59. S. B. Lee, H. M. Lee, M. Kiani, U. M. Jow, and M. Ghovanloo, "An inductively powered scalable 32-channel wireless neural recording system-on-a-chip for neuroscience applications," *IEEE Transactions on Biomedical Circuits and Systems*, vol. 4, pp. 360–371, 2010.
60. W. Xu, Z. Luo, and S. Sonkusale, "Fully digital BPSK demodulator and multilevel LSK back telemetry for biomedical implant transceivers," *IEEE Transactions on Circuits and Systems II: Express Briefs*, vol. 56, pp. 714–718, 2009.

References

61. M. Ghovanloo and S. Atluri, "A wide-band power-efficient inductive wireless link for implantable microelectronic devices using multiple carriers," *IEEE Transactions on Circuits and Systems I: Regular Papers*, vol. 54, pp. 2211–2221, 2007.
62. S. Atluri and M. Ghovanloo, "Design of a wideband power-efficient inductive wireless link for implantable biomedical devices using multiple carriers," in *2nd International IEEE EMBS Conference on Neural Engineering*, vol. 2005, (2005., Arlington, VA), pp. 533–537, 2nd International IEEE EMBS Conference on Neural Engineering, 2005.
63. G. Wang, W. Liu, M. Sivaprakasam, and G. A. Kendir, "Design and analysis of an adaptive transcutaneous power telemetry for biomedical implants," *IEEE Transactions on Circuits and Systems I: Regular Papers*, vol. 52, pp. 2109–2117, 2005.
64. Y. Hu and M. Sawan, "A fully integrated low-power BPSK demodulator for implantable medical devices," *IEEE Transactions on Circuits and Systems I: Regular Papers*, vol. 52, pp. 2552–2562, 2005. Dec.

2

System-Level Modeling

This chapter presents a comprehensive system-level modeling of an implanted wireless system with features including a wireless power and data transmission system and a closed-loop epileptic seizure detection and suppression device tailored for low-power implantable applications. The modeling process is executed utilizing the MATLAB Simulink software. The primary objective is to develop a robust system-level model that covers power delivery and control mechanisms, thermal dissipation and control strategies, as well as data transfer functionalities. This model serves as the foundation for the subsequent development of system and block-level specifications, post-integration of modules dedicated to power, temperature, and data aspects. The block diagram and overview of the proposed system are also presented. The system utilizes a single inductive link for power transmission and bidirectional data communication, integrated with a monitoring unit. It records high-resolution intracranial EEG (iEEG) signals via multiple electrode channels. The recorded signals are processed in the analog domain and subsequently fed into a digital signal processing block to extract five time-domain features for seizure onset detection, based on datasets from Inselspital Bern. Once a seizure is detected, a current-mode multi-channel electrical stimulator is activated, delivering electrical pulses to the brain as a therapeutic response.

2.1 System Overview

A closed-loop implantable medical device (IMD) includes two primary components: (1) data and power transmission systems and (2) stimulation and recording units [1, 2]. The external unit remotely supplies the necessary power, typically via magnetic coupling, eliminating the risk of infection associated with wired connections. Commands and recorded data are transmitted bidirectionally between the external and internal units. An analog front-end (AFE) block performs signal pre-processing in the analog domain, followed by a digital bio-signal processor that detects seizure onsets with high accuracy. Upon receiving a seizure detection flag from the digital signal processing (DSP) block, the neurostimulator temporarily delivers electrical pulses to the brain as a therapeutic response [3]. The effectiveness of a closed-loop implant heavily relies on the DSP's ability to detect seizures precisely and with minimal latency. Various feature extraction techniques have been reported, categorized into time-domain, frequency-domain, time-frequency domain, and nonlinear methods [3–5]. Current-mode neurostimulation is commonly employed to treat neurological conditions such as Parkinson's disease and epilepsy [6, 7]. Active and passive charge-balancing techniques are implemented to minimize residual charges at the electrode-electrolyte interfaces, thereby preventing damage to biological tissues and implanted electrodes.

Several complex scenarios may arise during wireless power and data transmission (WPDT) or stimulation, including coil misalignment, variations in distance, temperature elevation, insufficient or excessive power delivery, and memory saturation. Modeling and analyzing each system block are critical to identifying essential design parameters and addressing these challenges. A system-level model of a wirelessly powered closed-loop seizure detection device incorporates power delivery and control, data transfer, neural recording, and seizure detection systems. This model guides the development of block-level specifications, including modules for power management and data handling. Figure 2.1 shows the block diagram of system-level modeling of a wirelessly powered telemetry system.

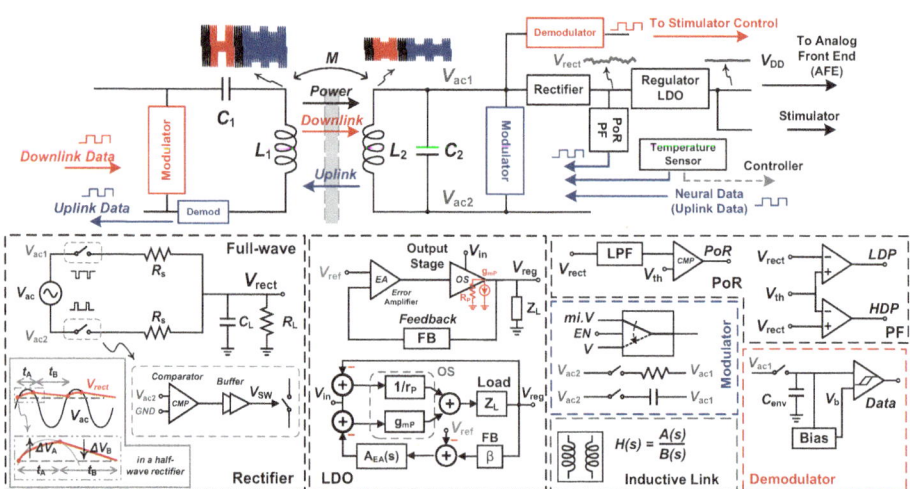

Fig. 2.1 Block diagram of system-level modeling of wirelessly powered telemetry system

2.2 Wireless Power System-Level Modeling

2.2.1 Full-Wave Active Rectifier

A rectifier is a nonlinear circuit that converts an AC input signal from an inductive link into an unregulated DC voltage, providing the supply voltage for IMDs. Rectifiers are generally categorized into two parts: (1) passive rectifiers, which utilize diodes for voltage conversion, and (2) active rectifiers, which employ comparators and switches for rectification (active diodes). The rectification process occurs when the input voltage exceeds the output voltage ($V_{in} \geq V_{out}$), at which point the switch is activated to charge the output capacitor. Conversely, when the input voltage falls below the output voltage ($V_{in} < V_{out}$), the switch is turned off, allowing the capacitor to discharge [8, 9]. The input voltage is typically modeled as a sinusoidal signal ($V_{in} = V_0 \sin(\omega_c t)$), delivered through the inductive link, while the output voltage is a DC level ($V_{out} = V_{rect}$), constrained by $V_{rect} \leq V_0$.

2.2.2 Low-Dropout Voltage Regulator (LDO)

The unregulated DC output voltage of the rectifier should be converted into a stable supply voltage using a voltage regulator, as the performance of the implant's electronic circuits is optimized for a specific DC supply level. Low drop-out (LDO) voltage regulators are widely discussed in the literature for this purpose. Voltage regulators typically consist of three essential parts: (1) a reference generator, (2) an error amplifier (EA), and (3) a pass device such as a power transistor. LDOs are categorized based on their pass transistors: (1) NMOS (2) PMOS pass transistors. In PMOS types (assuming the supply voltage is positive), LDOs utilize a common-source structure, while a source-follower should be used in NMOS types. The EA of the NMOS LDO regulator should be powered by a step-up converter to generate a sufficiently high voltage for driving M_N. Compared to a PMOS LDO, the NMOS LDO exhibits lower output impedance at frequencies exceeding the unity-gain bandwidth of the EA [10]. At such high frequencies, a PMOS LDO responds more slowly to load variations, relying solely on the load capacitor to supply a transient current. In contrast, for an NMOS LDO, when V_{out} decreases due to a sudden increase in load demand, M_N naturally experiences a higher gate-source voltage (V_{GS}), enabling it to promptly deliver additional current to the load. This behavior is an inherent characteristic of a source-follower stage. Additionally, since electrons in an N-type transistor exhibit higher mobility than holes in a P-type transistor, an NMOS LDO can be designed with a significantly smaller footprint, even when accounting for the area occupied by the additional step-up charge pump.

In an LDO regulator, the largest capacitances are typically the output filtering capacitor (C_L) and the parasitic gate capacitance (C_{Gate}) of the power MOS transistor. As a result, at least two low-frequency poles exist in the left-half-plane (LHP): one at the output node (p_{Out}) and another at the gate of the pass transistor (p_{Gate}). To achieve low transient ΔV_{out}

as low as possible, a dominant (p_{Out}) with an off-chip capacitor is chosen, as shown in Fig. 2.2. LDO's performances can be evaluated and measured by the following metrics:

- Dropout Voltage: $V_{in} - V_{out}$ at which V_{out} is no longer regulated. It depends on the pass device (PMOS) and the load current which is in the range of 100–500 mV.
- Quiescent current (I_Q): the consumed current of an LDO regardless of the load which consists of currents of the reference generator, error amplifier, feedback resistors, and support circuits. $I_Q = I_{in} - I_{out}$
- Efficiency (η) (current/power): can be defined based on (a) current and (b) power consumption as follows: $\eta_I = \frac{I_{out}}{I_{in}} = \frac{I_{out}}{I_{out}+I_Q}$, $\eta_P = \frac{I_{out} \cdot V_{out}}{(I_{out}+I_Q) \cdot V_{in}} \approx \frac{V_{out}}{V_{in}}$
- Voltage Regulation

    - Load regulation (static/dynamic): $\frac{\Delta V_{out}}{\Delta I_{out}}$ (mV/mA)
    - Line regulation (static/dynamic): $\frac{\Delta V_{out}}{\Delta V_{in}}$ (mV/V)

- Power Supply Rejection (PSR): regulator's ability to reject noise in V_{out} because of variation of power supply as an input voltage (V_{in}). It is similar to line regulation but is measured in the frequency domain and is similar to line transient but is measured for small signal variations. $PSR(f) = \frac{\Delta V_{out}(f)}{\Delta V_{in}(f)}$

According to Fig. 2.2, the system analysis and formulas can be derived using Mason's gain formula [11] for the transfer functions related to V_{in} and V_{ref}:

$$H_{in}(s) = \frac{V_{out}(s)}{V_{in}(s)}, H_{ref}(s) = \frac{V_{out}(s)}{V_{ref}(s)} \tag{2.1}$$

$$V_{out}(s) = H_{in}(s) \cdot V_{in}(s) + H_{ref}(s) \cdot V_{ref}(s) \tag{2.2}$$

$$\text{at DC: } Z_L = R_L, \ H_{EA}(s) = A_{EA0}, \ (R_{F1}+R_{F2})||R_L = R_{FL}, \ \beta = \frac{R_{F2}}{R_{F2}+R_{F1}} \tag{2.3}$$

where R_L, C_L, R_{ESR}, and $R_{F1,2}$ are the load resistor and capacitor, the equivalent series resistance of C_L, and feedback resistors, respectively. At DC bias:

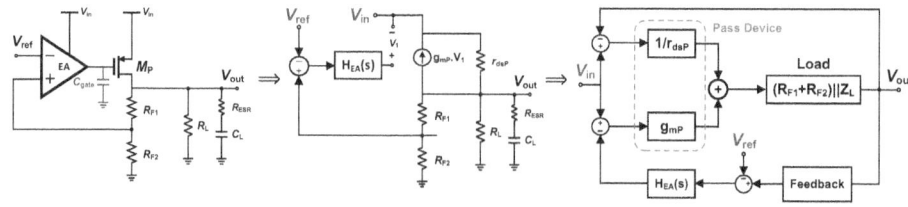

Fig. 2.2 Schematic of the system-level model of an LDO

$$H_{ref} = \frac{V_{out}}{V_{ref}} = \frac{A_{EA0} \cdot g_{mP} \cdot r_{dsP}}{1 + A_{EA0} \cdot g_{mP} \cdot r_{dsP} \cdot \beta + r_{dsP}/R_{FL}} \approx \frac{1}{\beta} \quad (2.4)$$

$$H_{in} = \frac{V_{out}}{V_{in}} = \frac{1 + g_{mP} \cdot r_{dsP}}{1 + A_{EA0} \cdot g_{mP} \cdot r_{dsP} \cdot \beta + r_{dsP}/R_{FL}} \approx \frac{1}{A_{EA0} \cdot \beta} \quad (2.5)$$

The line and load regulation (LNR and LDR) formula can be derived as:

$$\text{LNR:} \quad \frac{\Delta V_{out}}{\Delta V_{in}} = \frac{1 + g_{mP} \cdot r_{dsP}}{1 + A_{EA0} \cdot g_{mP} \cdot r_{dsP} \cdot \beta} \approx \frac{1}{A_{EA0} \cdot \beta} \Rightarrow \Delta V_{out} = \frac{\Delta V_{in}}{A_{EA0} \cdot \beta} + \frac{\Delta V_{ref} + \Delta V_{offset}}{\beta} \quad (2.6)$$

$$\text{LDR:} \quad \frac{\Delta V_{out}}{\Delta I_{out}} = \frac{r_{dsP} \| (R_{F1} + R_{F2})}{1 + \beta \cdot A_{EA0} \cdot g_{mP} \cdot [r_{dsP} \| (R_{F1} + R_{F2})]} \approx \frac{r_{dsP}}{1 + T_0} \quad (2.7)$$

where T_0 is the open-loop gain. Finally, these transfer functions can be implemented into the model.

2.2.3 Power Feedback and Power-on-Reset (PoR)

A power feedback system is designed to monitor the amount of power delivered to the IMD. If the rectified voltage surpasses a predefined threshold, the "high delivered power" (*HDP*) bit is activated. Conversely, if the voltage drops below the required level, the "low delivered power" (*LDP*) bit is activated [12]. The system-level model of the power feedback unit is illustrated in Fig. 2.1. The power-on reset (PoR), as depicted in Fig. 2.1, ensures that sufficient power is delivered before initiating data communication, thereby supporting robust telemetry.

2.3 Data Communication System-Level Modeling

Data communication paths connect the implantable medical devices and the external unit, facilitating forward (downlink) and backward (uplink) data transfer. Downlink transmission is used to send commands from the primary station to the devices, such as initiating a stimulation or recording phase. Uplink transmission sends biological parameters, such as neural recordings, or reports on the technical performance of the device, such as power consumption, to the external unit. In this system, data communication paths utilize the same magnetic field employed for remote powering, allowing for simultaneous power transfer and bidirectional half-duplex data communication.

2.3.1 Downlink Communication Model

Downlink data is modulated into the signal's amplitude using an amplitude shift keying (ASK) modulation scheme. A system-level model for ASK downlink communication consists of modulation and demodulation blocks. When $Data = 1$, the output signal remains unchanged; otherwise, the output is $MI \times Signal$, where MI is the modulation index ($0 \leq MI \leq 1$). For demodulation, the signal's envelope is extracted and converted to digital bits through an envelope detector and comparator [13]. To model an ASK demodulator and its transfer function, a simplified model of an inductive link used for data communication is shown in Fig. 2.3. The load at the implant is modeled as a pure resistor.

The frequency response of the inductive link is analyzed under various resonant configurations to identify the optimal resonant frequency for each implant [14]. The goal is to ensure that the difference between two adjacent resonant frequencies (ω_C) is significantly greater than the average of their bandwidths. The corresponding equations, as illustrated in Fig. 2.3, are provided below:

$$V_S = I_1 \cdot \frac{1}{sC_1} + V_1 + I_1 R_1 \quad V_1 = s.L_1.I_1 + s.M.I_2$$
$$V_{OUT} = I_2.R_2 + V_2 \quad\quad V_2 = s.M.I_1 + s.L_2.I_2 \quad (2.8)$$

where M is the mutual inductance, k represents the coupling coefficient ($M = k\sqrt{L_1.L_2}$), and L_1 and L_2 are the inductance values. The transfer function is:

$$H_{ASK}(j\omega) = \frac{V_{OUT}}{V_S} \quad (2.9)$$

Inductors L_1 and L_2 have a fixed inductance, and the capacitors, C_1 and C_2, are determined with respect to the resonant frequency (ω_C), $C_1 = \frac{1}{\omega_C^2 L_1}$ and $C_2 = \frac{1}{\omega_C^2 L_2}$. The simulation results of different resonant frequencies are shown in Fig. 2.4a. The resonant frequency ω_C sweeps from 5 to 20 MHz with a 2 MHz step. The critical coupling coefficient is $k_C^2 \approx \frac{R/R_2 + Q_{L_2}^2}{Q_{L_1} Q_{L_2} \sqrt{R^2/R_2^2 + Q_{L_2}^2}}$, where $Q_{L_{1,2}}$ are the quality factors of L_1 and L_2. When the

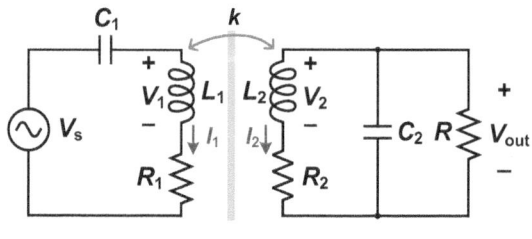

Fig. 2.3 Schematic of the inductive link

2.3 Data Communication System-Level Modeling

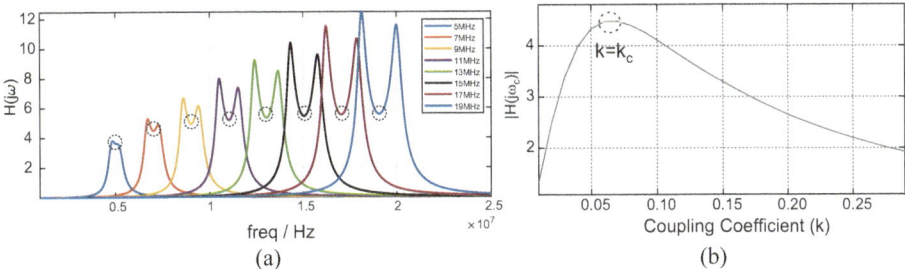

Fig. 2.4 Simulation results of the inductive link with the sweeping of **a** frequency and **b** coupling coefficient (k)

inductive link is strongly coupled ($k > k_C$), the frequency response exhibits two peaks near the resonant frequency. Conversely, when the coupling is weaker ($0 < k < k_C$), a single peak is observed at the resonant frequency, as shown, for example, in the 5 MHz plot in Fig. 2.4a. Furthermore, for strongly coupled coils, $|H_{ASK}(j\omega_C)|$ has nearly equal values at the resonant frequencies. At $\omega = \omega_C$, the amplitude response as a function of k can be approximated by the following expression:

$$|H_{ASK}(j\omega_C)|^2 \approx \begin{cases} \frac{L_2}{L_1 k^2} & \text{if } k > k_C \\ \frac{L_1 L_2}{R_1^2(\frac{L_2}{R}+\frac{R_2}{\omega_C^2 L_2})^2} k^2 & \text{if } 0 < k < k_C \end{cases} \quad (2.10)$$

The trend between $|H_{ASK}(j\omega_C)|$ and k is shown in Fig. 2.4b. In the implant units, the ASK demodulator operates within a specific amplitude range for the input signal. The desired implant is selected by tuning the carrier frequency to match the resonant frequency of its LC tank. Additionally, the voltage gain is highly dependent on the coupling coefficient (k), necessitating flexible adjustment of the transmitted signal amplitude to compensate for variations in k.

2.3.2 Uplink Data Communication Model

Data transmission in the uplink is achieved using the load shift keying (LSK) scheme, leveraging the same magnetic field used for power transfer. Modulation is performed by altering the resonance condition through a switch placed in parallel with the load, which is controlled by the data signal. Consequently, the voltages at the primary coil's terminals are modulated in response to the load variations, as illustrated in Fig. 2.1. The demodulator model comprises an envelope detector equipped with a bias generator and a decision unit, similar to the downlink demodulation process.

The model of an LSK circuit is depicted in Fig. 2.5. The LC tanks of the transmitter and receiver resonate at the same frequency when the targeted implant is selected. As illustrated

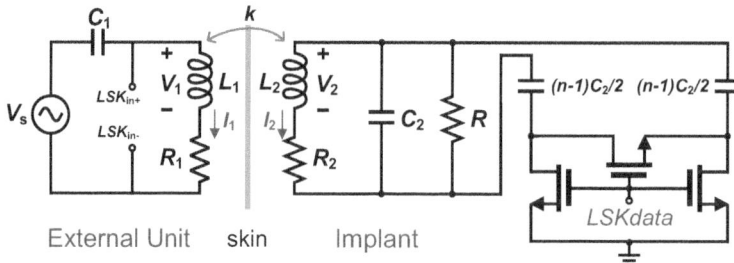

Fig. 2.5 Simplified model of LSK communication

in Fig. 2.5, three NMOS transistors function as switches to modulate data by controlling two capacitors connected in parallel with the receiver LC tank. When these capacitors are switched in parallel with C_2, the resonant condition is disrupted, thereby affecting the voltage at the transmitter ($V_{LSKin} = V_{LSKin+} - V_{LSKin-}$). In this state, the "LSKdata" signal is set to '1', and the equivalent receiver capacitance becomes $C_3 = C_2 + \Delta C = nC_2$. Conversely, when the capacitors are not engaged, the "LSKdata" signal remains '0', and the receiver capacitance remains unchanged. The voltage supplied to the ASK demodulator at the primary stage is given by $V_{LSKin} = V_{LSKin+} - V_{LSKin-}$. According to Fig. 2.5, the equations are expressed as follows:

$$\begin{cases} V_{LSKin} = V_1 + R_1.I_1 \\ (I_2.R_2 + V_2)\left(\frac{1}{R} + s.C_2\right) + I_2 = 0 \end{cases} \quad (2.11)$$

The LSK transfer function is:

$$H_{LSK}(j\omega) = \frac{V_{LSKin}}{V_S} \quad (2.12)$$

To validate the results, assuming $n = 1.5$, the amplitude responses of the uplink path for transmitting '0' and '1' are shown in Fig. 2.6. When transmitting '0', two peaks are observed, along with a valley at the resonant frequency, with a gain of approximately 3. For the transmission of '1', a prominent peak is present at the resonant frequency, with a gain exceeding 10. At higher resonant frequencies, the gain for '0' transmission decreases, while the gain for '1' transmission increases. Consequently, distinguishing between '0' and '1' becomes more efficient at higher resonant frequencies.

Further insights are obtained through the theoretical analysis presented below. The design of the LSK capacitor (C_3) must ensure that the transmitter's demodulator can reliably differentiate the voltage levels corresponding to '1' and '0' at the resonant frequency. By substituting $\omega = \omega_C$ under the resonant condition, the amplitude responses ($|H_{LSK'0'}(j\omega_C)|$ and $|H_{LSK'1'}(j\omega_C)|$) for '0' and '1' can be computed individually. The modulation index (MI) is then defined to verify the success of the reverse telemetry.

2.3 Data Communication System-Level Modeling

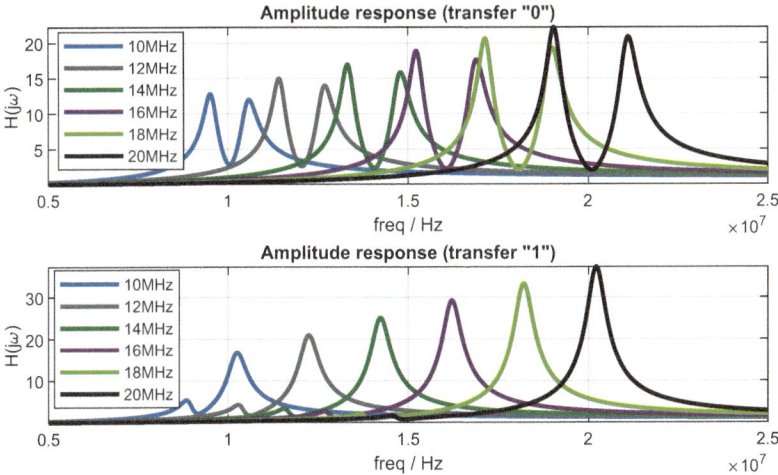

Fig. 2.6 Amplitude response of uplink LSK at different resonant frequencies (data '0' and '1' transmission) when $n = 1.5$

$$MI_{LSK} = \frac{|H_{LSK\cdot 1}(j\omega_C)| - |H_{LSK\cdot 0}(j\omega_C)|}{|H_{LSK\cdot 1}(j\omega_C)| + |H_{LSK\cdot 0}(j\omega_C)|} \quad (2.13)$$

The characteristics of the uplink LSK telemetry are illustrated in Fig. 2.7. The top-left subplot depicts the resonant gain variation with the resonant frequency when the uplink LSK transmits '0'. The gain decreases from approximately 3–2.3 as the frequency increases from 10 to 20 MHz, with a minimum gain of approximately 2.2 observed at a resonant frequency of 27 MHz. The bottom subplot shows the modulation index (MI) variation with n. Two potential solutions are proposed to achieve a high modulation index:

1. $n \geq 1.4$: This requires a capacitor in parallel with C_2, with a capacitance value greater than $0.4C_2$.
2. $n \leq 0.6$: This requires a capacitor in series with C_2, with a capacitance value less than $1.5C_2$.

The circuit shown in Fig. 2.5 adopts the first solution, incorporating two additional capacitors with at least $0.5C_2$ to achieve an MI exceeding 60%. The top-right subplot demonstrates the resonant gain variation at resonant frequencies when the uplink LSK transmits '1' under different additional parallel capacitance values. The block diagram of the wireless telemetry system is presented in Fig. 2.8. Power and data are transmitted simultaneously through the inductive links. The forward and backward data paths utilize ASK and LSK schemes, respectively. Each implant is equipped with an LC tank tuned to a specific resonant frequency, enabling the external device to select a particular implant by adjusting C_0 (a variable capacitor) in its LC tank.

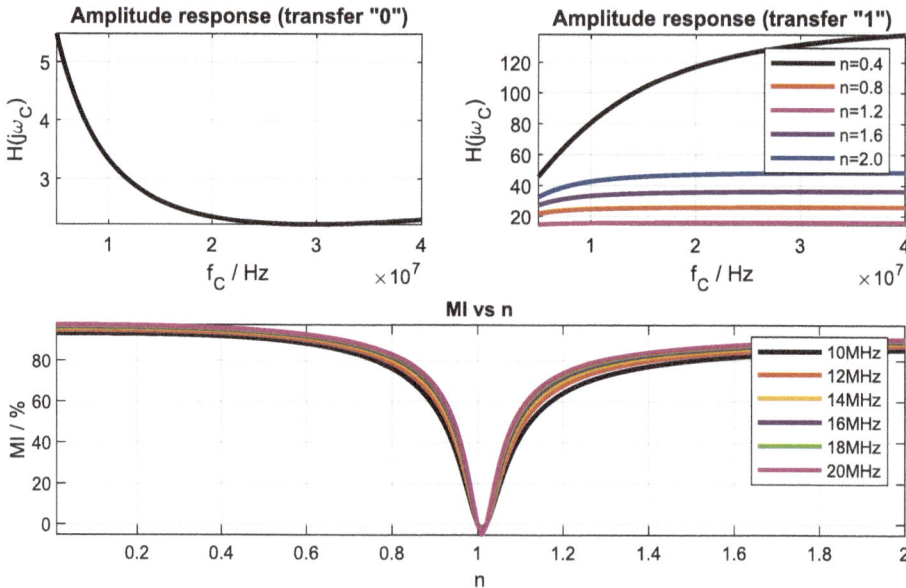

Fig. 2.7 Characteristics of uplink LSK telemetry

A simulation test bench has been developed to verify the system, consisting of one external device and three implants (implant 1 with $f_{c1} = 13.56$ MHz, implant 2 with $f_{c2} = 16$ MHz, and implant 3 with $f_{c3} = 20$ MHz). The simulation is divided into six phases: The first phase involves downlink data communication to a specific implant. The second and third phases are idle. The fourth to sixth phases involve uplink data communication, where data from each implant is sequentially sent back to the external unit. The simulation results for wireless telemetry with implant 1 ($f_{c1} = 13.56$ MHz) are shown in Fig. 2.9a. Each subfigure illustrates the received signal at each implant and the digital output data from its ASK demodulator. The LSK demodulator outputs digital data only during the fourth phase, as implant 1 sends data back to the external device in this phase. In the final subfigure, the red signal represents the downlink data, while the blue signal indicates the uplink data, demonstrating that only implant 1 communicates with the external device. Another example, depicted in Fig. 2.9b, adjusts the carrier frequency to 20 MHz, matching the resonant frequency of implant 3. In this scenario, only implant 3 transmits data (blue signal in the bottom subfigure) to the external device.

Finally, a system-level simulation is conducted for the entire WPDT system. Over a duration of $100\,\mu s$, the system transmits downlink data only during the interval of 20–$40\,\mu s$, as illustrated in Fig. 2.10a, b. Demodulating uplink data presents greater challenges compared to downlink data due to the low load resistance and high source resistance in reverse mode, resulting in minimal amplitude changes at the primary stage. Simulation results for the transmitted and demodulated downlink and uplink data, along with their envelope signals,

2.3 Data Communication System-Level Modeling

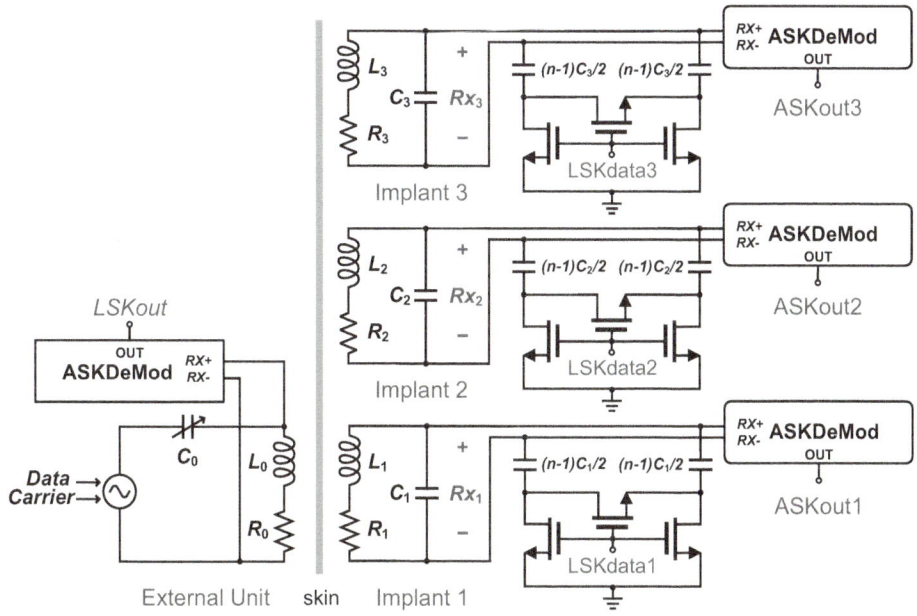

Fig. 2.8 Block diagram of the wireless telemetry system

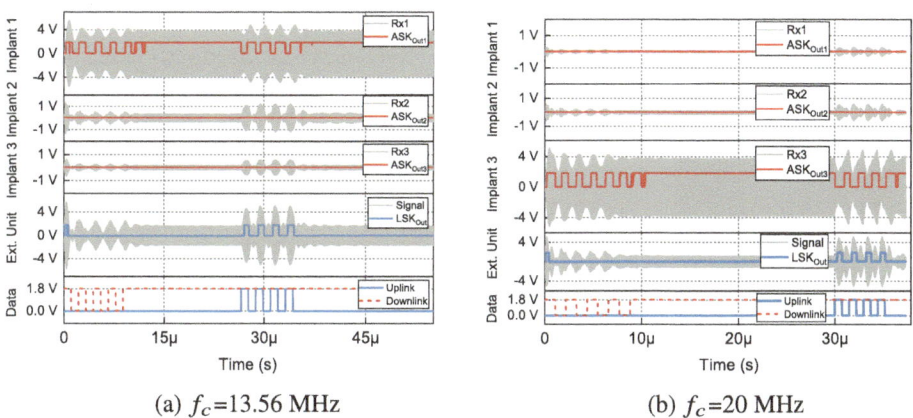

(a) f_c=13.56 MHz

(b) f_c=20 MHz

Fig. 2.9 Simulation results of the bidirectional data communication system to **a** implant 1 with f_{c1} = 13.56 MHz **b** implant 3 with f_{c3} = 20 MHz

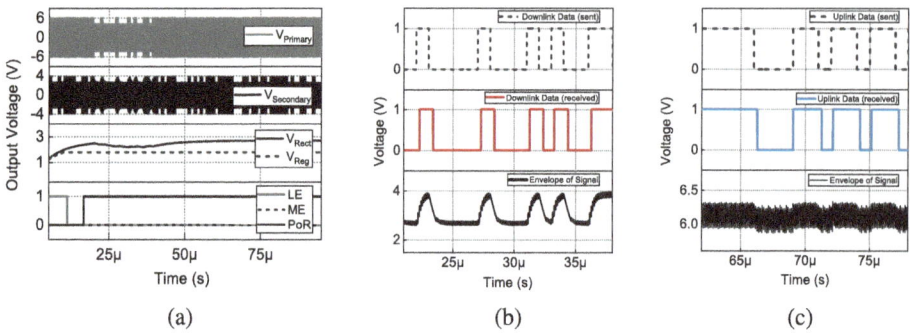

Fig. 2.10 Simulated results of **a** wireless power delivery, **b** downlink data, and **c** uplink data communication

Fig. 2.11 Flowchart of the closed-loop wireless implant (ME: more than enough and LE: less than enough wireless power condition)

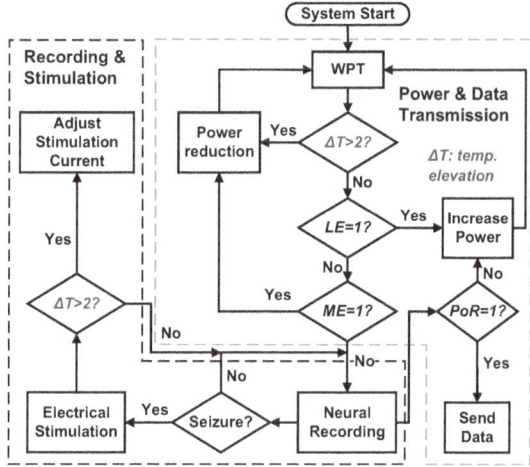

are shown in Fig. 2.10b, c. Figure 2.11 provides a flowchart depicting the system's behavior from initialization to stimulation. Critical parameters, including power-related variables and temperature elevation, are continuously monitored to ensure safe WPDT and stimulation without causing temperature elevation. With a 13.56 MHz input sine wave, the system provides stable voltage outputs at the rectifier (V_{rect}) and voltage regulator (V_{reg}), along with power control bits. A thermistor, modeled as a temperature sensor, measures temperature within the range of 25–40 °C and outputs a voltage corresponding to the sensed temperature.

2.4 Recording and Stimulation System Architecture

The proposed multi-channel epileptic seizure control implant includes an AFE for each channel, a compressive sensing data reduction module, an analog-to-digital converter (ADC), a digital signal processor (DSP), and an electrical neurostimulator with charge-balancing functionality. The block diagram of the system is illustrated in Fig. 2.12, and detailed descriptions of each block are provided in the subsequent sections.

2.4.1 Signal Pre-processing in the Analog Domain

The recording system of the implant includes an analog front-end (AFE) for each channel, consisting of two gain stages—a low-noise amplifier (LNA) and G_2—and a low-pass filter (LPF), as described in [3]. The AFE is responsible for amplifying the extremely weak neural signals within the frequency band of interest and canceling the large DC offset. In the Simulink model, the LNA is represented by a 14th-order Butterworth band-pass filter and a gain block, as detailed in [3]. For seizure detection applications, the bandwidth of interest is below 2 kHz, and a 17th-order analog Butterworth LPF is employed to implement the LPF block. Typical neural recorders provide approximately 45 dB amplification of iEEG signals, achieved through the LNA and G_2. A key consideration for seizure detection implants is reducing the input signal dimensions to minimize computational complexity in the digital domain. To this end, a compressive sensing data reduction (CSDR) block [3] is used to randomly select pre-processed analog signals from all available channels based on the system's compression ratio. To digitize the pre-processed, compressed iEEG signals, a 16-bit ADC is implemented using an ADC sample/hold and ADC quantizer in Simulink. The digitized signals are then delivered to a signal processing unit for further analysis.

2.4.2 Biomedical Digital Signal Processing

Digital signal processing (DSP) for epileptic seizure detection involves multiple feature extraction and classification steps. Five commonly used time-domain features are imple-

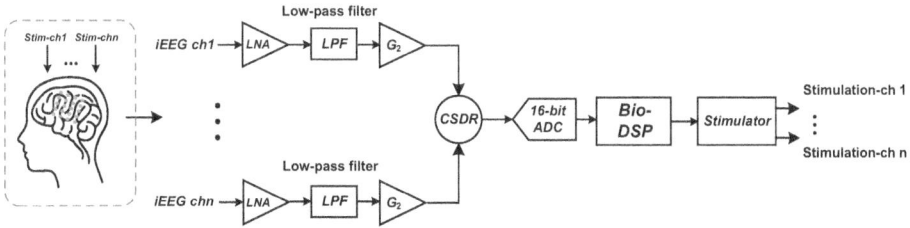

Fig. 2.12 Block diagram of an iEEG recording and seizure control system

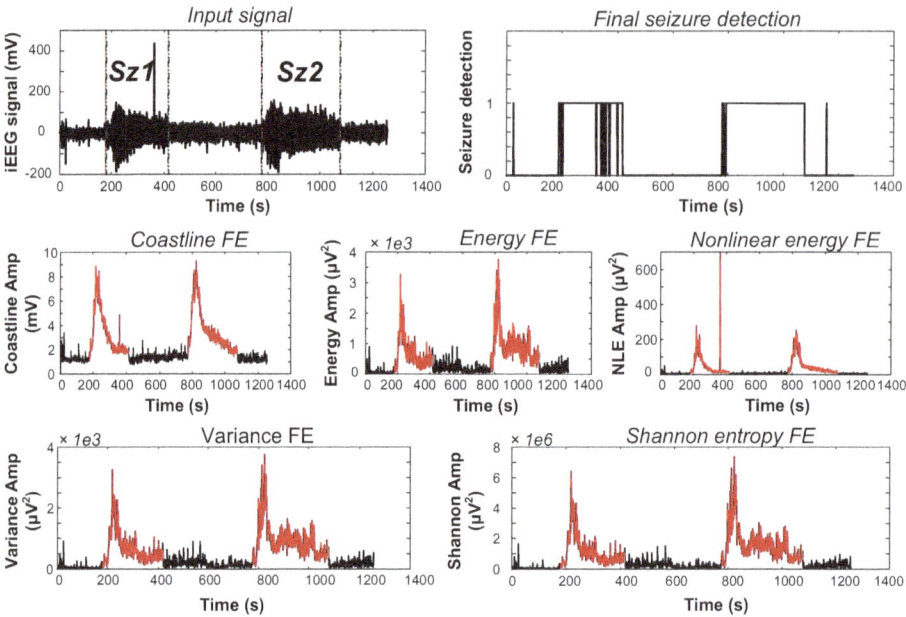

Fig. 2.13 Simulated results of feature extraction and seizure detection plots using datasets from Inselspital Bern [17]

mented in the Simulink model of the feature extractor [3, 15, 16]: (1) coastline, (2) energy, (3) nonlinear energy, (4) variance, and (5) Shannon entropy. Classification is performed based on threshold crossings, with optimized threshold values determined during the training phase to achieve the highest cross-validation detection accuracy. Additionally, a 3-s majority function post-processing method is employed to reduce false-positive detections [3]. The final seizure detection decision is made using a majority function with inputs from the five feature extractors. A detection flag is activated when at least three features indicate a seizure event. Figure 2.13 illustrates the input iEEG signal from a patient containing two seizure events, the outputs of the five feature extractor cores, and the final seizure detection signal generated by the DSP block. The iEEG signals used for signal processing simulations are sourced from the SWEC-ETHZ database, collected at Inselspital Bern [17].

2.4.3 Neural Stimulator and Charge Balancer

A biphasic current-mode neural stimulator is modeled in MATLAB Simulink, integrating a 5-bit current digital-to-analog converter (DAC) to control the amplitude of cathodic and anodic pulses. The system delivers a peak current of 1 mA, operating at a 1 kHz stimulation frequency with a pulse width of $100\,\mu s$. Inherent mismatches between cathodic and

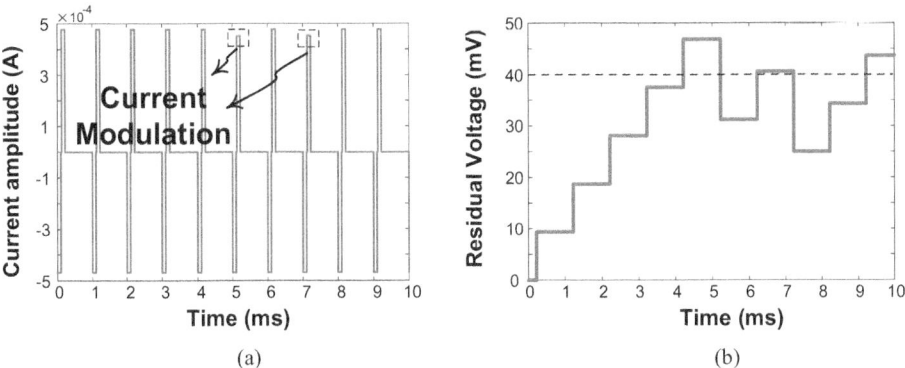

Fig. 2.14 a Biphasic current stimulation, and **b** residual voltage

anodic pulse amplitudes lead to residual voltage buildup at the electrode-tissue interface. To counteract this, an anodic current modulation-based charge balancer is employed to regulate excessive residual voltage. Activation occurs when the residual voltage surpasses a 40 mV safety threshold, adjusting the anodic current to restore voltage levels to a safe range. Figure 2.14 illustrates the stimulation waveform and residual voltage variations. The residual voltage exceeds 40 mV after five, seven, and nine stimulation cycles, prompting the charge balancer to dynamically modulate the anodic current, ensuring voltage stabilization in subsequent cycles.

2.5 Summary

This chapter introduces a comprehensive system-level model for a WPDT system tailored for multi-channel iEEG signal recording and processing. The system incorporates compression and conversion of signals from the analog to the digital domain. Each WPDT unit is meticulously analyzed and represented as a system-level block. Seizure detection is achieved through feature extraction techniques, including coastline, energy, nonlinear energy, variance, and Shannon entropy. During seizure events, a multi-channel biphasic current-mode electrical stimulator is employed to deliver periodic pulses to the brain, while an integrated charge-balancing mechanism ensures that residual charges are managed effectively to prevent potential tissue damage.

References

1. M. J. Karimi, K. Farhang Razi, C. Dehollain, and A. Schmid, "Modeling and Analysis of a Wirelessly Powered Closed-Loop Implant for Epilepsy," in *2022 IEEE Biomedical Circuits and Systems Conference (BioCAS)*, pp. 414–418, 2022.
2. M. J. Karimi, M. Jin, Y. Zhou, C. Dehollain, and A. Schmid, "Wirelessly Powered and Bi-directional Data Communication System with Adaptive Conversion Chain for Multisite Biomedical Implants Over Single Inductive Link," *IEEE Transactions on Biomedical Circuits and Systems*, pp. 1–11, 2024.
3. K. F. Razi, M. J. Karimi, C. Dehollain, and A. Schmid, "System-Level Modeling of a Safe Autonomous Closed-loop Epileptic Seizure Control Implant," in *2021 4th International Conference on Bio-Engineering for Smart Technologies (BioSMART)*, pp. 1–4, 2021.
4. K. F. Razi and A. Schmid, "Two-stage Hardware-Friendly Epileptic Seizure Detection Method with a Dynamic Feature Selection," in *2021 43rd Annual International Conference of the IEEE Engineering in Medicine Biology Society (EMBC)*, pp. 156–159, 2021.
5. Z. Jiang and W. Zhao, "Optimal Selection of Customized Features for Implementing Seizure Detection in Wearable Electroencephalography Sensor," *IEEE Sensors Journal*, vol. 20, no. 21, pp. 12941–12949, 2020.
6. J.-Y. Son and H.-K. Cha, "An Implantable Neural Stimulator IC With Anodic Current Pulse Modulation Based Active Charge Balancing," *IEEE Access*, vol. 8, pp. 136449–136458, 2020.
7. A. H. Hassan, Z. E. Mohamed, A. E. Fahmy, H. Mostafa, and A. M. Soliman, "Design Trade-Offs for Neural Stimulators Optimization," in *2020 IEEE International Symposium on Circuits and Systems (ISCAS)*, pp. 1–5, 2020.
8. G. Bawa and M. Ghovanloo, "Analysis, design, and implementation of a high-efficiency full-wave rectifier in standard CMOS technology," *Analog Integrated Circuits and Signal Processing*, vol. 60, pp. 71–81, 2009.
9. M. J. Karimi, C. Dehollain, and A. Schmid, "An offset-enhanced active rectifier with delay compensated active diodes for wirelessly powered biomedical implants," *Electronics Letters*, vol. 60, no. 15, p. e13274, 2024.
10. Y. Lu and W.-H. Ki, *CMOS Integrated Circuit Design for Wireless Power Transfer*. Singapore: Springer, 2018.
11. S. J. Mason, "Feedback Theory-Some Properties of Signal Flow Graphs," *Proceedings of the IRE*, vol. 41, no. 9, pp. 1144–1156, 1953.
12. M. J. Karimi, C. Dehollain, and A. Schmid, "Power Feedback Control Unit for Closed-Loop Wirelessly Powered Biomedical Implants," *IEEE Transactions on Circuits and Systems II: Express Briefs*, vol. 70, no. 5, pp. 1674–1678, 2023.
13. M. J. Karimi, Y. Zhou, C. Dehollain, and A. Schmid, "An Analysis of An ASK Demodulator With Dual Self-Biased Separated Voltages for Implantable Applications," *IEEE Transactions on Circuits and Systems II: Express Briefs*, pp. 1–1, 2024.
14. M. J. Karimi, Y. Zhou, C. Dehollain, and A. Schmid, "Simultaneous Wireless Power and Data Transmission Through a Single Inductive Link For Multiple Implantable Medical Devices," in *2022 17th Conference on Ph.D Research in Microelectronics and Electronics (PRIME)*, pp. 329–332, 2022.

References

15. P. Boonyakitanont, A. Lek-uthai, K. Chomtho, and J. Songsiri, "A review of feature extraction and performance evaluation in epileptic seizure detection using EEG," *Biomedical Signal Processing and Control*, vol. 57, p. 101702, 2020.
16. K. F. Razi and A. Schmid, "Epileptic Seizure Detection with Patient-Specific Feature and Channel Selection for Low-power Applications," *IEEE Transactions on Biomedical Circuits and Systems*, pp. 1–10, 2022.
17. A. Burrello, L. Cavigelli, K. Schindler, L. Benini, and A. Rahimi, "Laelaps: An energy-efficient seizure detection algorithm from long-term human ieeg recordings without false alarms," in *2019 Design, Automation & Test in Europe Conference & Exhibition (DATE)*, pp. 752–757, 2019.

Implantable Dual-Band Inductive Link 3

This chapter explores dual-band telemetry systems and presents a dual-band inductive link operating at 6.78 and 13.56 MHz within the industrial, scientific, and medical (ISM) frequency bands. The primary goal is to enable wireless power and data transmission for multisite biomedical implants. To minimize magnetic interference between coils and optimize the area, a single-coil transmitter configuration is adopted in the presented design. Additionally, this chapter examines the benefits of using various planar coil shapes, including square, circular, and octagonal topologies. The proposed link ensures a reliable power supply for electronic circuits while supporting downlink data communication.

3.1   General Context

In wireless power transmission (WPT), dual-band communication has demonstrated significant potential [1–6]. Dual-band operation allows systems to function at two distinct frequencies, enhancing flexibility and efficiency. This methodology offers several advantages, such as increased bandwidth, reduced crosstalk, improved power transfer efficiency by utilizing power at two frequencies, and minimized network congestion through distributed information across bands. It supports parallel communication and enhances the system's overall robustness. Existing research has delved into dual-band systems, particularly focusing on the impact of coil shape selection [7–11]. While these studies have contributed valuable insights, they are often confined to conventional coil shapes, such as circular or square designs [7, 9], or specific applications like electric vehicle charging [8, 10]. A systematic investigation into close-field applications, coupled with wireless power and data delivery tests, remains essential. Simulation-based studies, while insightful, face inherent limitations. The complexity of dual-band systems, interactions between multiple parameters, and the physical intricacies of coil design are often oversimplified in simulations. These simplifications can lead to incomplete or potentially misleading conclusions, warranting caution in their interpretation and application. To overcome these limitations, practical implementations—both at the board and silicon levels—are necessary to validate findings and establish robust conclusions.

Addressing these gaps, this chapter proposes a dual-band transmitter and receiver coil configuration specifically designed for wireless power and data transmission. To assess the impact of different coil shapes on system performance, finite element modeling (FEM) is employed as a numerical analysis technique. The primary objective of this chapter is to provide a realistic and detailed understanding of dual-band coil design, with a particular focus on its application in biomedical implants.

3.2 Dual-Band Coil Designing Method

The circuit diagram of the presented wireless power and data transmission (WPDT) system is shown in Fig. 3.1, utilizing dual-band coils operating at 13.56 and 6.78 MHz. The external unit sends ASK modulated data modulated onto a power waveform. The implanted units, tuned to different resonance frequencies, receive both power and data, extracting the wireless power and demodulating the data. Each implant consists of three main components: (1) an inductive link, (2) a wireless power conversion unit, and (3) a data demodulation unit. The inductive link serves as the core of the system, facilitating energy transmission and reception through magnetic fields. The wireless power management unit generates the necessary voltage and current for the electronic circuitry, employing an active rectifier, voltage reference, and low-dropout voltage regulator. The data unit extracts both data and clock signals from the received waveform.

This chapter focuses on the design of the dual-band coil receiver and transmitter. Electromagnetic simulations are performed using Ansys Electronic Desktop. The Maxwell 3D Design module [12] is employed for modeling the coil, which is then integrated with the rest

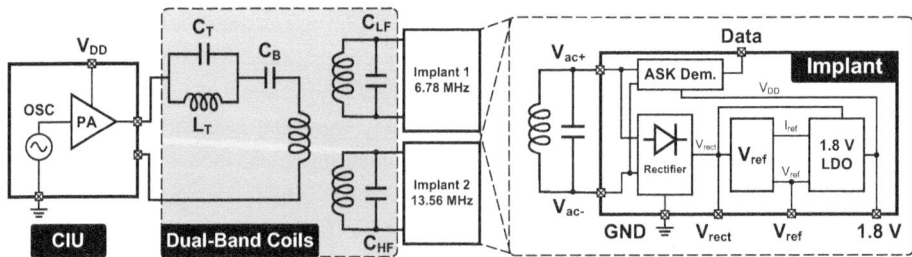

Fig. 3.1 Circuit diagram of the presented dual-band inductive link used in a multisite wireless power and data transmission system. The system operates at 6.78 and 13.65 MHz

3.2 Dual-Band Coil Designing Method

of the circuit using Ansys Twin Builder (Simplorer) [13] to analyze its frequency domain behavior. Key design parameters investigated include the shape of the coil (e.g., square, octagonal, or circular) and the spatial configuration of the transmitter (Tx) and receiver (Rx), as these directly influence the performance of wireless power transfer coils. Performance metrics evaluated include coil area, quality factor (Q), self-resonance frequency (SRF), and coupling coefficient (k). The design optimized for the highest power transfer efficiency (PTE) is selected for fabrication and subsequent validation experiments.

The Tx and Rx coils are subject to size constraints determined by the requirements for brain implantation. The outer diameter of the Tx coil must range from 24 to 43 mm, while the Rx coil has the range between 5.5 and 13.5 mm. These constraints are specified at the start of the investigation. The dual-band transmission frequencies are fixed at 6.78 and 13.56 MHz, as these frequencies are internationally reserved for ISM applications. The separation distance between the Tx and Rx coils varies from 5 to 10 mm. Simulations are conducted in both vacuum and brain environments, with the latter characterized by a relative permittivity ϵ_r of 170 and a bulk conductivity of 0.141 S/m, representing the properties of brain white matter over the frequency range of 6.78–13.56 MHz. This approach allows for a comparative analysis of inductances and coupling coefficients between the two environments. The planar coil shapes significantly influence the coupling between the Tx and Rx coils. Factors such as the angle at each coil turn impact parasitic resistance and, consequently, the Q-factor. The shape of the Tx magnetic field also affects the allowable separation between the coils. Simulations are performed for circular, square, and octagonal Tx and Rx coils in various combinations, where the two Rx coils in each configuration share the same design. The inductances of the Tx (L_{Tx}) and Rx (L_{Rx}) are fixed at $4 \pm 0.02\,\mu H$ and $0.3 \pm 0.02\,\mu H$, respectively. The total occupied area (A_{occ}), defined as the area of the smallest square enclosing the entire coil, is fixed at 1600 mm^2 for the Tx coil and 144 mm^2 for the Rx coil. Under these size constraints, the octagonal Rx coil closely resembles the circular Rx coil and is therefore excluded from further analysis. The coil thickness is set at 1.0 oz of copper (35 μm). Detailed geometric parameters for the coils are presented in Table 3.1. The CAD model of the coil is created using Ansys's in-built RectHelix geometry configurator. The model is then simulated using the Maxwell 3D Design tool, with parasitic resistance computed from coil ohmic losses via a field calculator.

Depending on the spatial placement of the coils relative to one another, the PTE can vary. The closer and more aligned the Tx and Rx coils are, the higher the coupling coefficient becomes. In a typical configuration, the two receiver coils are positioned on the same plane as the Tx coil, with a separation distance of $D_{TxRx} = 5$ mm (referred to as the parallel configuration). The center-to-center distance between the two Rx coils, denoted as D_{Rx}, is shown in Fig. 3.2. The maximum spatial coverage of the Tx-Rx combination can be determined by varying D_{Rx}. A second coil configuration, referred to as the overlapped configuration, is also examined. In this scenario, the two Rx coils are center-aligned with the Tx coil, as shown in Fig. 3.2. One Rx coil is placed closer to the Tx, while the second is positioned $D_{TxRx} = 5$ mm away from the first Rx coil. The distance between the first Rx coil and the

Table 3.1 Geometry of the designed coils [11]

Coil	Shape	A_{in}^a (mm²)	Width/Spacing (mm)	N_{turn}	R_{par}^b (LF/HF) (Ω)
Tx1	Circle	236.7	0.48/0.48	10	2.5/3.4
Tx2	Square	750.8	0.35/0.5	8	3/4
Tx3	Octagon	330.2	0.35/0.5	9	2.8/3.4
Rx1	Circle	19.22	0.3/0.3	5	0.59/0.78
Rx2	Square	67.24	0.25/0.3	4	0.67/0.87

[a] Area of the biggest square that can fill the inner coil space
[b] Parasitic coil resistance at 6.78 MHz (LF) and 13.56 MHz (HF)

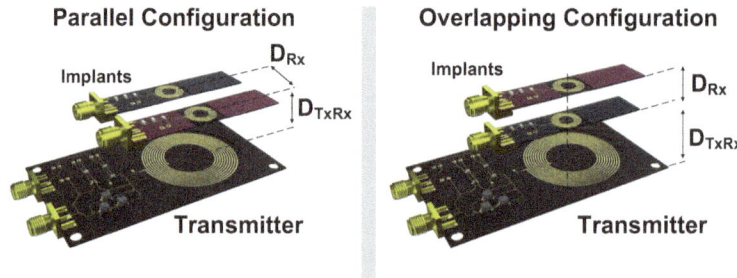

Fig. 3.2 Spatial configuration of the Tx and Rx coils in parallel and overlapped configurations

Tx is denoted as D_{TxRx}. The system is designed to operate at two resonance frequencies. The Tx coil is connected to a capacitor, forming an LC branch, which is subsequently connected in series to an LC tank [2], as shown in Fig. 3.3. When the source signal operates at 6.78 MHz, the frequency is below the LC tank's natural resonance frequency, causing the tank to behave inductively. This inductive behavior enables the LC branch to resonate at 6.78 MHz (f_{LF}). At 13.56 MHz, the LC tank behaves capacitively, allowing the LC branch to resonate at this higher frequency (f_{HF}). A trap frequency is selected between these two resonance frequencies. The relationship between the three frequencies (f_{LF}, f_{HF}, f_{trap}) is expressed as:

$$f_{HF} = \sqrt{2} f_{trap} = 2 f_{LF} \tag{3.1}$$

In the receiver circuit, the Rx coil is connected in parallel with a passive capacitor and a resistive load to reduce output impedance for effective load driving. The capacitor also tunes the Rx coil to the desired receiving frequency. To simulate the parasitic resistance of the source and coils, a virtual resistor is added to the circuit. The simulated parasitic resistances of the Tx and Rx coils (R_{Tx} and R_{Rx}) are provided in the last column of Table 3.1. A parametric sweep is conducted to determine the optimal capacitance for the highest voltage gain at the specified frequencies with a resistive load of 100 Ω. Once the capacitance is

3.3 Measurement Results

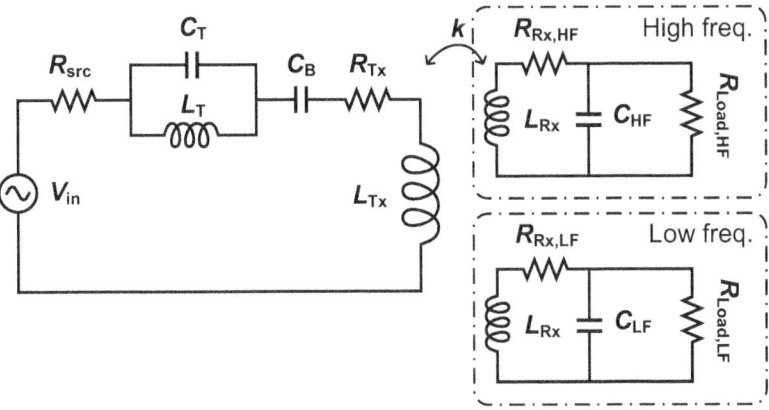

Fig. 3.3 Circuit model of the dual-band inductive link [11]

Table 3.2 Circuit parameters

Index	Description	Value	Unit
C_B	Branch capacitance	68.9	pF
C_T	Tank capacitance	137.9	pF
C_{LF}	LF capacitance	1.6	nF
C_{HF}	HF capacitance	410	pF
L_T	Tank inductance	2	μH
R_{src}	Source resistance	50	Ω

defined, another parametric sweep is performed to identify the optimal resistive load for maximum PTE. The selected values for the passive components are presented in Table 3.2.

3.3 Measurement Results

Multiturn coils in various shapes are fabricated on 1 mm thick FR4 substrate printed circuit boards (PCBs). The primary coil characteristics are measured using a vector network analyzer (VNA). A 3D-printed rack is designed and manufactured to securely hold the PCB at varying distances and configurations during the experiment, as illustrated in Figs. 3.4 and 3.5. Both parallel and overlapped configurations are tested with different combinations of Rx and Tx coils. In the parallel configuration, the distance $D_{Tx\text{-}Rx}$ is fixed at 5 mm, and measurements are taken while varying D_{Rx}. Conversely, in the overlapped configuration, D_{Rx} is fixed at 5 mm, and measurements are taken while varying $D_{Tx\text{-}Rx}$. The configurations achieving the best coupling coefficient are then selected for the PTE experiment. The PTE is derived from S_{12} parameters [14]. The designed coils exhibit parasitic capacitance

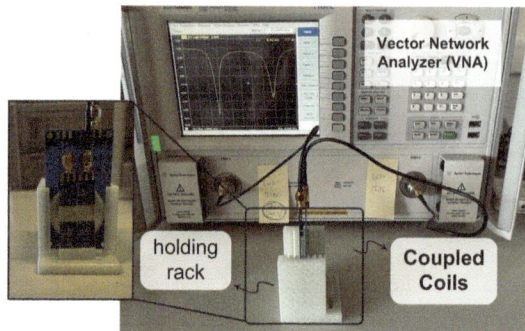

Fig. 3.4 Coil measurement setup comprising a vector network analyzer (VNA), SMA cables, a holding rack, and the coil PCB. This configuration is employed to evaluate the inductance and Q-factor of the coils. For dual-band communication measurements, the direct-to-coil port is linked to the left terminal of the VNA, while the measured coil is connected to the right terminal. The PCB integrates impedance matching circuits, with one of the receiver coils shorted during measurements

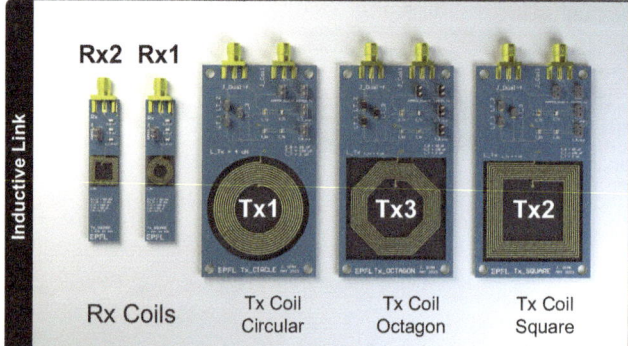

Fig. 3.5 Measurement setup of the dual-band inductive link and fabricated coils in different geometries

at the pF level, resulting in SRFs of 12 GHz for the Tx coils and 70 GHz for the Rx coils. These SRFs are significantly higher than the operating frequency of 13.56 MHz, allowing the parasitic capacitance to be neglected. The Q-factor is calculated using resistance values from Table 3.1, showing that square coils have lower Q-factors compared to circular coils. However, the Q-factor alone does not fully determine the final PTE. The experimental results are summarized in Table 3.3. The measured inductances are slightly higher than the design values.

3.3 Measurement Results

Table 3.3 Inductance and quality factor measurements

Tx/Rx	Inductance (μH)	HF Q-factor	LF Q-factor
Tx1	5.19	11.4	36.2
Tx2	5.34	9.18	33.0
Tx3	5.34	9.95	34.0
Rx1	0.421	28.3	22.1
Rx2	0.425	26.1	19.9

3.3.1 Parallel Configuration

Six Rx-Tx coil combinations are analyzed, with their simulated coupling coefficients presented in Fig. 3.6a. As expected, all coefficients exhibit a quadratic decline as separation distance increases. When the receiving coils are closely spaced (<20 mm), the Tx1-Rx2 combination achieves the highest coupling coefficient. For intermediate separations (>21 mm), Tx3-Rx2 outperforms the others, while at larger distances (above 26.5 mm), Tx2-Rx2 provides the highest coupling. Thus, the optimal coil selection depends on the specific scenario. Among all configurations, Tx3-Rx2 (octagonal Tx with square Rx) demonstrates the highest average coupling coefficient across both short and large separations. Experimental results in Fig. 3.6b confirm the simulations, following a parabolic decrease as D_{Rx} increases. Two distinct coupling coefficient groups emerge experimentally, though at slightly higher D_{Rx}. The Tx3-Rx2 combination remains the most effective for parallel configuration transmission. Additionally, Fig. 3.7 presents the measured S-parameters of the designed coils in this configuration.

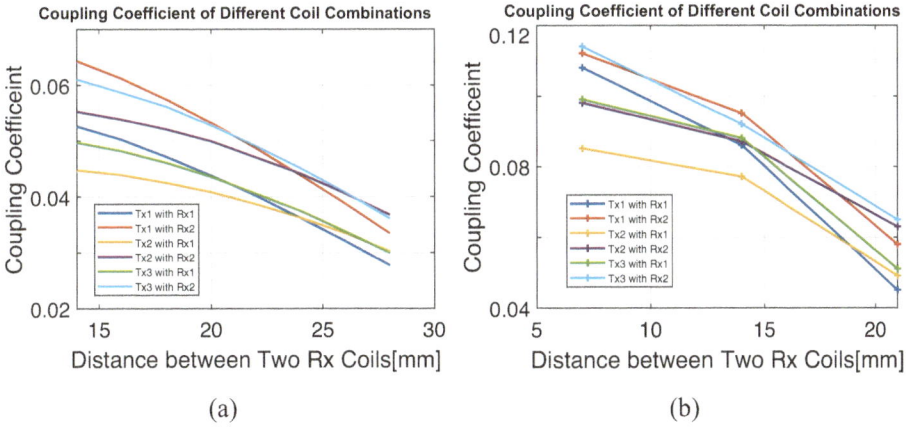

Fig. 3.6 **a** Simulated and **b** measured results of coupling coefficients for various combinations and configurations

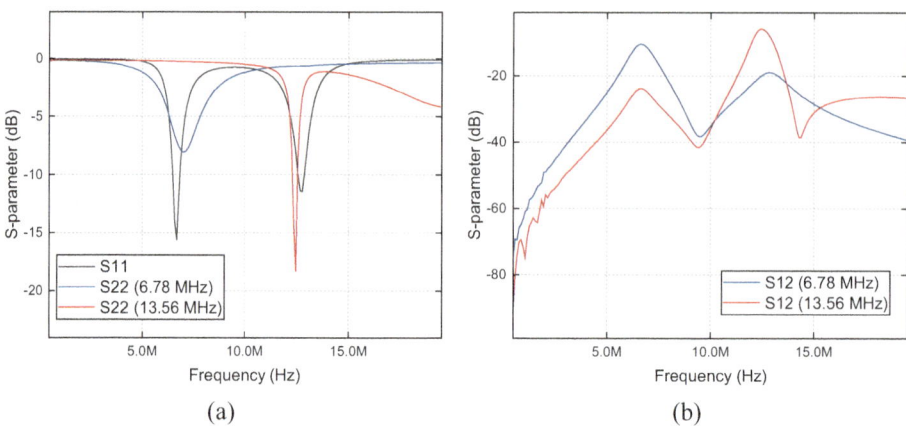

Fig. 3.7 Measured results of the S-parameter of the Tx3-Rx2 coils in the parallel configuration

3.3.2 Overlapping Configuration

Twelve placement combinations are simulated to study how coil shapes affect the coupling coefficient. The Rx coils are positioned directly beneath the Tx coil, with the closer Rx referred to as Coil1 and the farther Rx as Coil2. The distance between Coil1 and Coil2 is fixed at 5 mm, while the distance between Tx and Coil1 varies from 2.5 to 7.5 mm. The measurement results [11] indicate that square Rx coils provide better coupling coefficients with the Tx coil compared to circular Rx coils. Additionally, circular Tx coils yield higher coupling coefficients overall than octagonal or square Tx coils. Circular Tx coils achieve coupling coefficients of 0.115 and 0.145 with circular and square Rx coils, respectively. To ensure even excitation of both receivers from the Tx, the difference between their coupling coefficients (Δk) should be minimized. Positioning a circular Rx closer to the Tx and a square Rx farther away, with a square Tx, achieves an average $\Delta k = 0.005$ at the cost of reducing the mean coupling coefficient to $\bar{k} = 0.062$. A better solution involves using an octagonal Tx, which results in $\Delta k = 0.012$ and $\bar{k} = 0.074$. Experiments in [11] conducted over a longer distance range compared to simulations reveal a hyperbolic decline in the coupling coefficient, rather than the parabolic decrease observed in simulations. Despite this difference, placing the square Rx farther from the Tx still reduces Δk. The shape of the Tx coil has a less pronounced effect in this scenario. The minimum average Δk occurs with a square Tx ($\Delta k = 0.016$), while the octagonal Tx achieves the highest $\bar{k} = 0.074$.

3.3.3 Frequency Response and PTE

Frequency responses are examined in both simulations and experiments to identify the optimal coil shape combinations for maximizing PTE. Frequency sweeps are performed

3.4 Summary

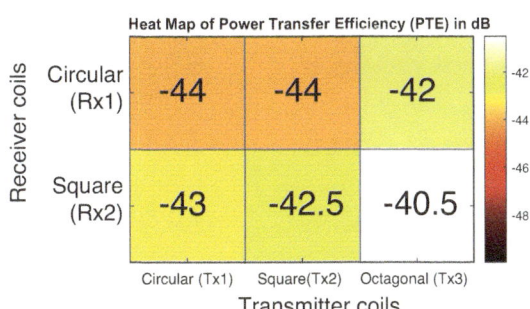

Fig. 3.8 Heatmap of the average PTE in the parallel configuration, showing that the octagonal Tx coil achieves the highest average PTE [11]

for both parallel and overlapped configurations. The PTE bode diagram exhibits the typical summit-and-valley shape, with summits at dual-band resonance frequencies and valleys at the trap frequency. A notable observation is that the PTE at the lower frequency band is approximately 10 dB higher than at the higher frequency band, a phenomenon not observed in simulations and indicative of greater impedance mismatch in the high-frequency band. As discussed, the Tx3-Rx2 configuration (octagonal Tx with square Rx) delivers the best performance over a broad range. Measurement results confirm that Tx3-Rx2 achieves a higher average PTE than Tx2-Rx2. Figure 3.8 presents a heatmap showing the average PTE for different Tx-Rx combinations in parallel configurations. In the overlapped configuration, experiments are conducted for D_{Tx-Rx} ranging from 5 to 20 mm, with D_{Rx} fixed at 5 mm. Placing the high-frequency Rx farther from the Tx leads to a drastic drop in its PTE, with a difference of approximately 20 dB. To achieve a balanced PTE between the two Rx coils, the high-frequency Rx should always be positioned closer to the Tx. For longer distance ranges, octagonal Tx coils are compared with circular Tx coils. Results, visualized in Fig. 3.9a, b, indicate that the Rx2-Rx2 receiver orientation achieves the highest PTE with Tx1 but shows significant differences between the two coils. Conversely, the Rx2-Rx1 orientation with Tx3 minimizes the differences between the two Rx coils.

3.4 Summary

This chapter presented the design of a dual-band inductive link for wireless power and data transmission, focusing on the impact of coil shapes on coupling performance. Circular, square, and octagonal Tx spiral coils were analyzed. The octagonal Tx coil combines the strengths of circular and square coils, making it particularly suitable for distance-sensitive applications like wireless communication in IMDs. Experimental results demonstrate that the octagonal Tx coil supports a broader physical range of signal transmission while reducing PTE differences caused by minor movements. Optimization of wireless coupling through

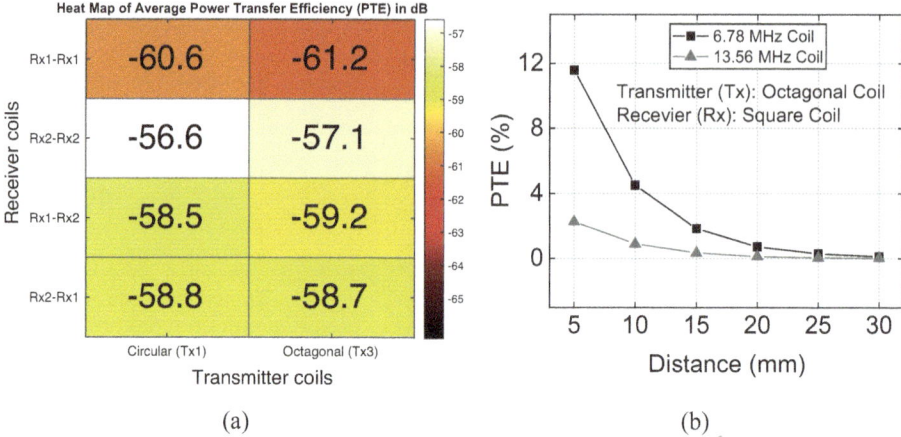

Fig. 3.9 Measured PTE heatmap. Rx1-Rx2 denotes the configuration where Rx1 is positioned closer to the transmitter, while Rx2 is placed further away. **a** Heatmap of the average PTE value in dB for both receivers in the overlapped configuration. **b** Measured PTE across different Tx-Rx coil distances [11]

experimentation highlights the benefits of using octagonal Tx coils with square Rx coils in parallel configurations. For overlapped configurations, arranging Rx coils based on resonance frequency minimizes PTE disparities, further improving system performance.

References

1. M. J. Karimi, A. Schmid, and C. Dehollain, "Wireless Power and Data Transmission for Implanted Devices via Inductive Links: A Systematic Review," *IEEE Sensors Journal*, vol. 21, no. 6, pp. 7145–7161, 2021.
2. M.-L. Kung and K.-H. Lin, "Enhanced Analysis and Design Method of Dual-Band Coil Module for Near-Field Wireless Power Transfer Systems," *Microwave Theory and Techniques, IEEE Transactions on*, vol. 63, pp. 821–832, 03 2015.
3. M. J. Karimi, M. Jin, Y. Zhou, C. Dehollain, and A. Schmid, "Wirelessly Powered and Bi-directional Data Communication System with Adaptive Conversion Chain for Multisite Biomedical Implants Over Single Inductive Link," *IEEE Transactions on Biomedical Circuits and Systems*, pp. 1–11, 2024.
4. M.-L. Kung and K.-H. Lin, "Dual-Band Coil Module With Repeaters for Diverse Wireless Power Transfer Applications," *IEEE Transactions on Microwave Theory and Techniques*, vol. 66, no. 1, pp. 332–345, 2018.
5. I. Habibagahi, R. P. Mathews, A. Ray, and A. Babakhani, "Design and Implementation of Multisite Stimulation System Using a Double-Tuned Transmitter Coil and Miniaturized Implants," *IEEE Microwave and Wireless Technology Letters*, vol. 33, no. 3, pp. 351–354, 2023.

References

6. M. J. Karimi, Y. Zhou, C. Dehollain, and A. Schmid, "Simultaneous Wireless Power and Data Transmission Through a Single Inductive Link For Multiple Implantable Medical Devices," in *2022 17th Conference on Ph.D Research in Microelectronics and Electronics (PRIME)*, pp. 329–332, 2022.
7. X. Shi, C. Qi, M. Qu, S. Ye, G. Wang, L. Sun, and Z. Yu, "Effects of coil shapes on wireless power transfer via magnetic resonance coupling," *Journal of Electromagnetic Waves and Applications*, vol. 28, no. 11, pp. 1316–1324, 2014.
8. T. Bouanou, H. El Fadil, A. Lassioui, O. Assaddiki, and S. Njili, "Analysis of Coil Parameters and Comparison of Circular, Rectangular, and Hexagonal Coils Used in WPT System for Electric Vehicle Charging," *World Electric Vehicle Journal*, vol. 12, no. 1, 2021.
9. M. McDonough and B. Fahimi, "Comparison between circular and square coils for use in Wireless Power Transmission," in *9th IET International Conference on Computation in Electromagnetics (CEM 2014)*, pp. 1–2, 2014.
10. E. Aydin, Y. Kosesoy, E. Yildiriz, and M. Timur Aydemir, "Comparison of Hexagonal and Square Coils for Use in Wireless Charging of Electric Vehicle Battery," in *2018 International Symposium on Electronics and Telecommunications (ISETC)*, pp. 1–4, 2018.
11. M. J. Karimi, J. Qian, C. Dehollain, and A. Schmid, "Design of a Dual-Band Wireless Power and Data Transfer Coil for Multisite Biomedical Implants," in *2023 IEEE Nordic Circuits and Systems Conference (NorCAS)*, pp. 1–6, 2023.
12. "Ansys Maxwell | Electromechanical Device Analysis Software — ansys.com." https://www.ansys.com/products/electronics/ansys-maxwell. [Accessed 02-2025].
13. "Ansys Twin Builder | Create and Deploy Digital Twin Models — ansys.com." https://www.ansys.com/products/digital-twin/ansys-twin-builder. [Accessed 02-2025].
14. Z. Despotovic, D. Reljic, V. Vasic, and D. Oros, "Power Transfer Analysis of an Asymmetric Wireless Transmission System Using the Scattering Parameters," *Electronics*, vol. 10, no. 8, 2021.

Wireless Power Conversion Chain and Control Methods

4

This chapter presents the remote powering and conversion mechanisms applied to wireless cortical implants, and more specifically developing the three-layer architecture explained in Chap. 1. The central implanted unit (CIU), implanted internally, serves as the primary wireless power source for the implant (ASPs), with wireless power delivery achieved through magnetic coupling and the received power must be converted into a stable DC voltage. The design of power conversion units is presented in this chapter. Moreover, developing the power control unit and the automatic resonance tuning system are presented. The design of the wireless power transfer units for the CIU unit is also proposed. The power dissipated inside the implant leads to a temperature elevation in the brain, which may disrupt the normal functioning of the brain. The temperature elevation of the brain implants should be less than 1 °C according to safety regulations [1, 2]. This temperature elevation equals a power density of 40 mW/cm^2 [3]. The maximum power consumption should be limited to 10 mW to ensure the safe operation of the implanted system [4].

4.1 System Overview

The power conversion chain in wirelessly powered IMDs includes different parts [5], as shown in Fig. 4.1. A full-wave active rectifier converts the AC voltage received from the inductive link to a DC voltage. This rectified voltage is further regulated and stabilized by designing a voltage reference and a low dropout (LDO) voltage regulator to minimize the supply voltage tolerances that affect the performance of the implant. In addition, the control units of the IMD are developed. The power feedback unit (PF) monitors the power delivered to the implant to ensure that the received voltage remains within acceptable operational

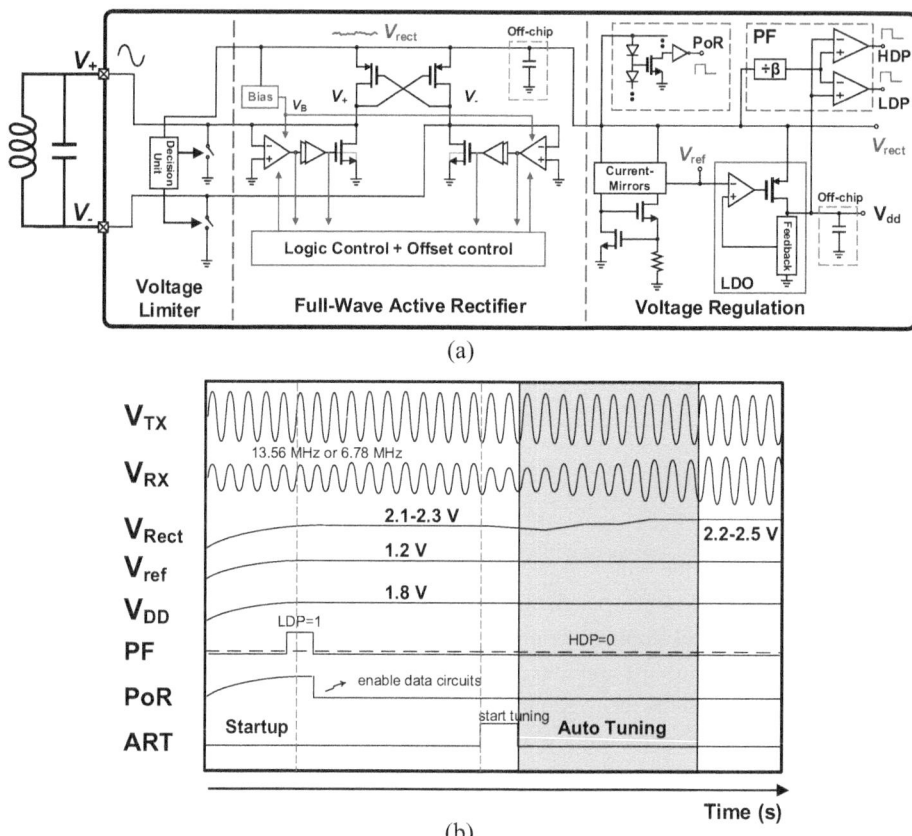

Fig. 4.1 a Block diagram of the wireless power conversion chain (PCC) and control units, **b** timing diagram of the proposed system

parameters. The power-on-reset (PoR) unit initializes the data communication after receiving sufficient power. The voltage limiter (VL) protects the implant against overvoltages. Finally, the automatic resonance tuning (ART) system optimizes wireless power transfer efficiency by maintaining an optimal coupling between the CIU and ASPs. The timing diagram of the system behavior is shown in Fig. 4.1b. Initially, the sine wave is received by the implanted coil. After a start-up phase, the rectifier voltage is stabilized, and the LDO and reference voltages are supplied. Meanwhile, the PF unit monitors the received voltage by generating two control bits. After a certain period, the ART system can be activated to optimize the resonance capacitor. After a certain time for the tuning process, the maximum rectified voltage is achieved. It should be noted that the VL unit continuously monitors potential overvoltages.

4.2 Full-Wave Active Rectifier

A wireless power transfer unit is developed to transfer power from the CIU to the implants. An on-chip oscillator generates the clock required for the magnetic field. A power amplifier (PA) transmits the power signal to the inductive link. Finally, a DC-DC converter adjusts the battery voltage level to align with the required voltage supply of the CIU electronic circuitry.

4.2 Full-Wave Active Rectifier

The implanted coil receives an AC voltage that needs to be converted into a DC voltage by the rectifier. This converted DC voltage will then serve as the power supply for the system. Full-wave active rectifiers are widely developed to convert the AC signal, as shown in Fig. 4.2. Full-wave rectification is preferred since it provides a more constant DC voltage than half-wave rectification. Also, active rectifiers offer higher conversion efficiency than passive rectifiers by implementing active diodes (a switch with a comparator) to minimize the drop-out voltages across the diodes and to maximize the voltage conversion ratio (VCR), the ratio of the rectified voltage to the amplitude of the input voltage ($V_{rect}/|V_{ac}|$).

Delays in the comparators of active rectifiers can lead to late turn-off of active diodes, causing them to remain ON even when V_{ac} falls below V_{rect}. As a result, reverse current flows through the circuit, ultimately reducing its power conversion efficiency (PCE). Various design techniques have been proposed to mitigate these switching delays in active diodes. One approach involves incorporating a switch to control the reverse current within the comparator [6, 7]. However, this method has certain limitations, including reduced conduction time and sensitivity to process variations. Another technique employs a time-varying or V_{out}-dependent offset [8], which introduces negative feedback to dynamically adjust the comparator offset based on V_{rect}. Additionally, an adaptive delay time control strategy, as proposed in [9], enables near-optimal on- and off-delay compensation. A more effective approach to mitigating comparator delays involves adaptively controlling its offset using negative feedback tailored to the circuit's operating conditions. This method compensates

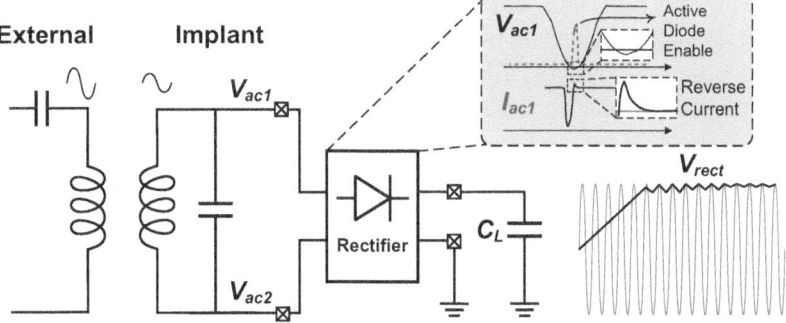

Fig. 4.2 Block diagram of an active rectifier connected to an inductive link in the IMDs

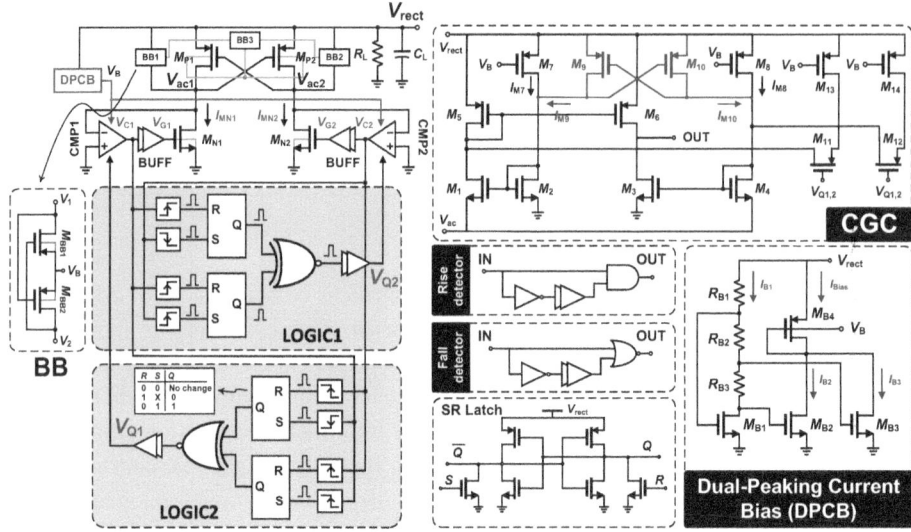

Fig. 4.3 Schematic of the presented full-wave digitally controlled active rectifier, along with the logic control unit, which includes fall and rise-edge detectors and an SR-latch

for delays while accounting for process variations. However, applying this technique to both turn-on and turn-off delay reduction requires stability analysis and may shorten conduction time. Alternative solutions, such as those presented in [10, 11], leverage high-speed comparators to minimize circuit delays.

The schematic of the proposed digitally-assisted active rectifier is illustrated in Fig. 4.3. The design consists of a PMOS pair ($M_{P1,2}$), NMOS active diodes ($M_{N1,2}$), and common-gate comparators (CGCs) CMP1 and CMP2 [5, 6, 12], along with their associated digital blocks, including an adaptive ON/OFF transition calibrator and three bulk biasing units (BBs). The transistor dimensions are optimized to minimize voltage drops across transistors while achieving an optimal balance between high VCR and PCE. To ensure optimal layout compatibility and maintain a differential topology, a 3-Dynamic BB (DBB) structure is employed. In this configuration, two DBBs are used to select the highest voltage between $V_{ac1,2}$ and V_{rect}, while the third DBB determines the highest voltage among the two, enabling a symmetric and common-centroid layout for $M_{P1,2}$.

4.2.1 Bias Circuit

For a robust and stable design, the active rectifier must initiate operation before any subsequent circuits can function. Therefore, the biasing circuit should be self-starting and independent of an external start-up circuit. A dual-peaking current biasing (DPCB) circuit is

4.2 Full-Wave Active Rectifier

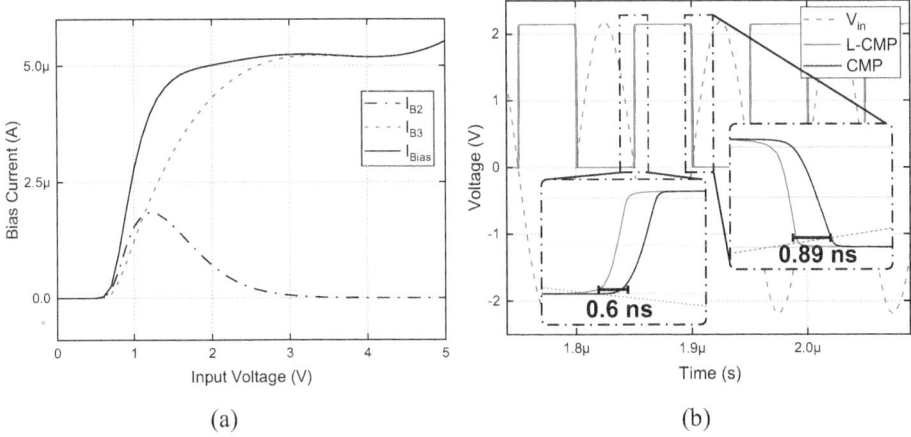

Fig. 4.4 Simulated results of the **a** generated bias currents of the DPCB and **b** the comparator improvement in the latched structure (L-CMP)

implemented [8], as illustrated in the DPCB box in Fig. 4.3. In this circuit, resistors R_{B2} and R_{B3} define two distinct peak current points for the output currents $I_{B2,3}$. These currents are then merged through transistor M_{B4}. The simulation results of the DPCB circuit are presented in Fig. 4.4a.

4.2.2 Offset-Controlled Common-Gate Comparator

The CGCs, depicted in the upper right box of Fig. 4.3, are widely used in active rectifiers due to their inherently low input resistance. However, their turn-on/off delay and the charging time required for the large NMOS switch capacitors can degrade the rectifier performance. To mitigate this issue, increasing the bias current $I_{M7,8}$ is a common approach, but it results in higher power consumption. To reduce turn-on/off delays, the current difference between I_3 and I_6 is increased by introducing cross-coupled transistors $M_{9,10}$ [10]. Lowering V_{ac} increases both the current and gate-source voltage of M_1. Simultaneously, a decrease in V_{ac} reduces the gate voltage of M4, increasing its source-gate voltage and consequently boosting the current of M_9. This process raises the gate voltage of M_2, enhancing the current through M_1 and, consequently, M_6. This occurs because M_1's source voltage drops while its gate voltage increases. Similarly, the gate voltage of M_2 increases, reducing the current and gate-source voltage of M_4 by decreasing the current through M_{10}. As a result, the comparator's output current $(I_6 - I_3)$ is increased, thereby improving switching speed. Increasing the sizes of $M_{9,10}$ introduces hysteresis in the comparator, requiring a recovery period. Thus, the dimensions of $M_{9,10}$ must be carefully optimized. The simulation results comparing a conventional comparator (CMP) and a latched comparator (L-CMP), showing 0.6 ns and 0.89 ns improvements in on/off delay, are presented in Fig. 4.4b.

4.2.3 Logic Unit

The gray boxes in Fig. 4.3 illustrate the schematics of the logic circuits integrated into the design. The logic unit consists of falling and rising edge detectors, SR latches, XNOR gates, inverters, and buffers to generate control signals and offsets [13]. Specifically, in LOGIC1, when V_{C1} in CMP1 decreases and VC2 in CMP2 increases, the falling and rising edge detectors generate a high output, setting V_{Q2} to zero. This mechanism injects current and applies an offset into CMP2 through transistors $M_{11,12}$, thereby optimizing comparator delay and extending conduction time.

4.2.4 Measurement Results

The presented rectifier is designed and fabricated in a 180 nm standard CMOS technology with a total area of 0.12 mm × 0.264 mm, as shown in Fig. 4.5. The implemented board contains an off-chip 2.5 nF capacitor. The rectifier's performance across various output power levels is assessed by sweeping the load resistor (R_L), as shown in Fig. 4.6a, which shows the trend of VCR and PCE of the rectifier. The PCE measurement is conducted by incorporating a 10 Ω resistor in series with the rectifier. The input power is measured using two differential active probes—one to capture the input voltage and another to measure the current through the series resistor. The parasitic capacitors at the input of the rectifier during large dI/dt transients are the main reason that the measured PCE is lower than the simulated one for low values of the load resistor [14]. Figure 4.6b presents the measured output of the rectifier. The design achieves the maximum VCR and PCE of 93 and 80.8%. Finally, Table 4.1 provides a design benchmark. This design presents efficient active rectification, excelling in terms of PCE and VCR when compared to similar designs. However, using an on-chip output capacitor facilitates system-on-chip integration, yet it comes at the cost of consuming a substantial amount of silicon area.

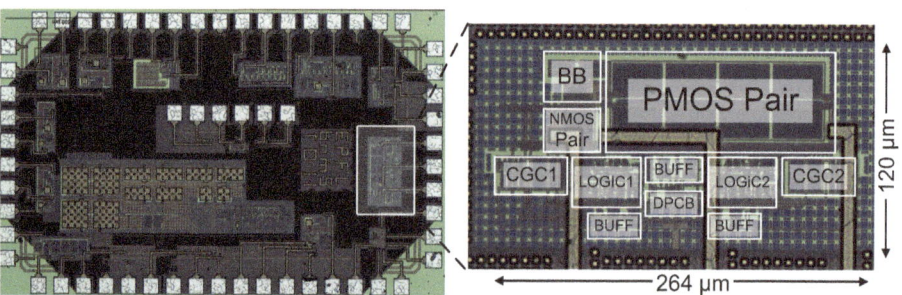

Fig. 4.5 Die photograph of the presented active rectifier fabricated using a standard 180 nm CMOS process, occupying an area of 0.12 mm × 0.264 mm

4.3 Voltage Reference

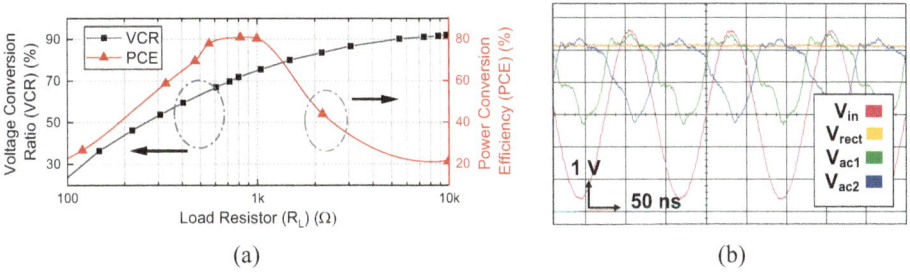

Fig. 4.6 Measured results of the **a** PCE and VCR across various loads and **b** transient input and rectified voltages

Table 4.1 Benchmark of active rectifiers [12]

	TVLSI'18 [15]	JSSC'19 [9]	JSSC'20 [16]	TCAS-I'20 [17]	TBCAS'21 [11]	JSSC'23 [18]	This work
Technology (μm)	0.35	0.18	0.18	0.18	0.18	0.18	0.18
Area (mm^2)	0.38	0.117	0.853	0.208	0.479	1.488	0.036
Frequency (MHz)	13.56	13.56	1–10	40.68	6.78	40.68	13.56
Load capacitor (nF)	1.7 nF (on-chip)	47 nF (off-chip)	0.13 n 1.2 u (off-chip)	2 nF (on-chip)	4.5 nF (on-chip)	1.8 nF (on-chip)	2.5 nF (off-chip)
Load resistor (kΩ)	0.5	0.51	0.3	0.5	0.5	0.5	0.5
Input voltage (V)	2.9–5.4	1–2.5	1.8–5	2.5–4	2.5–5	1.5–4	1.5–4
Max. P$_{out}$ (mW)	126.7	34.1	231.6	56.6	125	24.8	61.25
Max. VCR (%)	92.7	94.9	88.6	84.1	95.4	93	93
PCE (%)	84.6–86.1	85.0–94.1	84.4–91.5	70.7–80.9	92.7–95.0	81.9–86.0	80.2–80.8
Method	Adaptive on/off	Adaptive on/off	Adaptive on/off	Adaptive on/off	Adaptive on/off	Adaptive on/off	Adaptive on/off + CGC improve
Compensation	Voltage	Delay	Current	Voltage + delay	Voltage + delay	Delay	Delay

4.3 Voltage Reference

Designing a stable and efficient voltage reference for LDO regulators in wirelessly powered biomedical implants presents a crucial challenge. These devices demand stringent power management and precision in voltage regulation to ensure the reliable operation of sensitive integrated circuits while minimizing energy consumption and voltage variation through supply voltage changes. The voltage reference plays a pivotal role in this context by providing a stable reference voltage to the LDO regulator, which, in turn, regulates the implant's operating voltage. This section presents the design of a 1.2 V voltage reference, which utilizes a fully CMOS implementation. Subsequently, simulation and measurement results are provided.

4.3.1 Circuit Design

The schematic of the voltage reference is presented in Fig. 4.7, which comprises a current source and a bias voltage circuit to generate the reference voltage, and startup. Assuming that the size of M_1 is large enough, $V_{GS1} \simeq V_{th1}$, the voltage at the resistor terminal (R_1) generates a reference current ($\simeq V_{th1}/R_1$). Therefore, a current based on the threshold voltage is created. Then, this current is mirrored to the output stage and generates reference currents to feed the comparators or operational transconductance amplifiers (OTAs). This current is also mirrored to the output stage and generates a reference voltage at the drain of M_{R1}, operating in the sub-threshold region. The voltage generation employs diode-connected transistors to minimize the area, while consuming low power. The output voltage is 1.2 V with a variation of 0.46 mV/V per input voltage variation and a power consumption of 53 µW at a 1.8 V voltage supply. DC input sweeping with measured and simulated results of the reference voltage are shown in Fig. 4.8a. Monte Carlo simulation results of the voltage

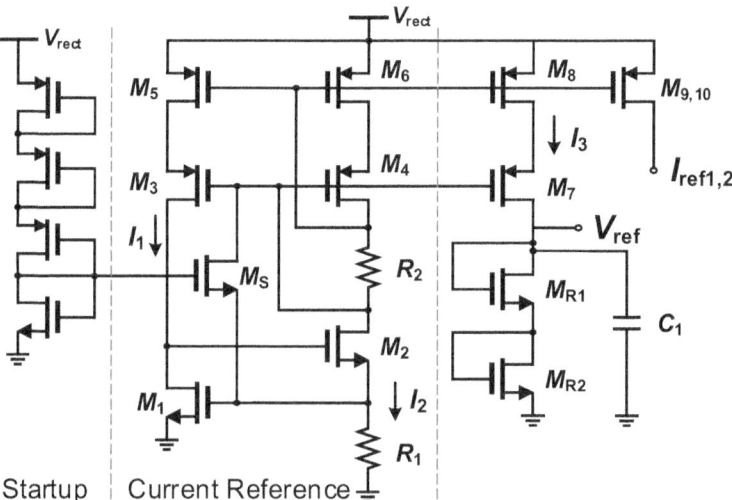

Fig. 4.7 Schematic of the voltage and current reference generator

4.4 Low-Dropout (LDO) Voltage Regulator

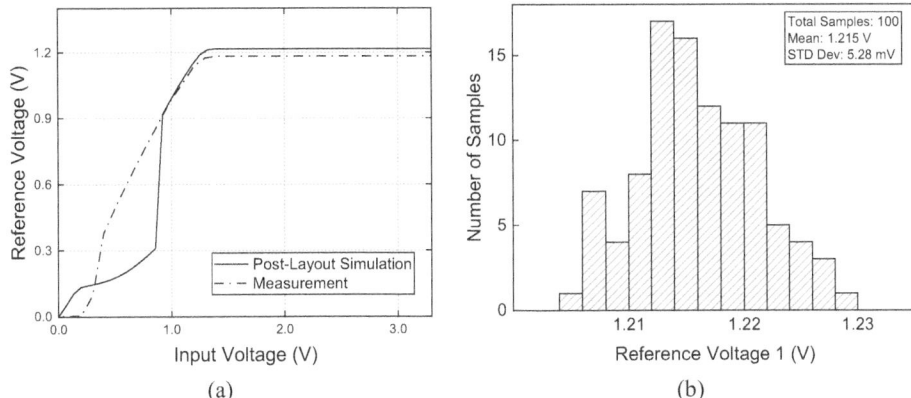

Fig. 4.8 Simulation and measurement results of CMOS voltage reference V_{ref1} by **a** input voltage sweeping and **b** Monte Carlo simulation results of the output reference voltage

reference are shown in Fig. 4.8b with 100 samples and a standard deviation of 5.28 mV (0.4% of the designed value). This variation is attributed to changes in voltage thresholds resulting from process variations.

4.3.2 Measurement Results

The presented voltage reference is fabricated in a 180 nm standard CMOS technology with a total area of 0.125 mm × 0.062 mm. Figure 4.8a displays the measured waveforms that emphasize the performance of the reference voltage at different input DC voltages. Figure 4.9 also shows the variation in V_{ref}, revealing overshoot and undershoot of 64 mV and 69 mV, respectively, when the input voltage changes from 2 V to 2.5 V.

4.4 Low-Dropout (LDO) Voltage Regulator

The unregulated DC voltage of the rectifier should be regulated to a stable DC voltage supply by a voltage regulator since the performances of the implant are optimized for a certain DC supply voltage. Moreover, the effects of coil misalignment and small ripples of the output of the rectifier can be filtered by LDOs. An LDO consists of three parts: an error amplifier (EA), a pass transistor, and a feedback unit. The design of two LDOs is presented in the following: (1) 1.8 V LDO with an off-chip output capacitor and (2) 1.2 V LDO with capacitor-less output.

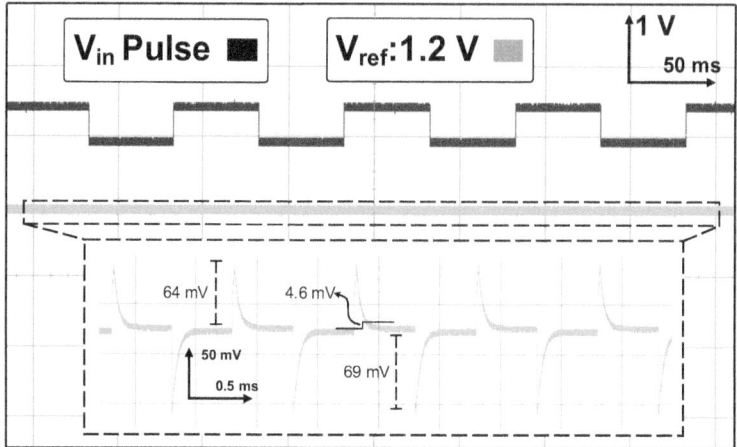

Fig. 4.9 Measurement of the variation of the reference voltage obtained by changing the input voltage with the zoomed overshoot and undershoot of the output voltage

4.4.1 1.8 V LDO with Off-Chip Output Capacitor

An LDO must be designed to provide a stable 1.8 V supply voltage. The important design parameters of the LDO are:

- Output voltage: 1.8 V, input voltage: 2–2.5 V (input reference: 1.2 V)
- Load current: from 10 μA to 10 mA, max. quiescent current: 20 μA
- Output capacitor: 1–10 μF

For an input voltage from 2 to 2.5 V with an output voltage of 1.8 V the allowable dropout voltage is between 200 and 700 mV. To bias the PMOS pass transistor in the saturation region throughout the maximum load current, it is essential to carefully consider an appropriately high W/L dimension. However, it is important to note that increasing the dimensions significantly leads to a larger parasitic gate capacitance. A buffer is designed between the pass transistor and the EA to control the stability of the LDO, as shown in Fig. 4.10. This technique addresses two low-frequency poles caused by the large gate capacitance of the pass transistor. The designed buffer must have small input capacitance, small output impedance, and large voltage swing. As calculated in the system-level modeling (2.7) and (2.6), since both line and load regulation are inversely dependent on the error amplifier DC gain, a higher gain offers better line/load regulation.

The schematic of the proposed LDOs is presented in Fig. 4.11. The reference voltage (V_{ref}) is derived from a designed voltage reference that is set at 1.2 V. This design offers a higher loop gain and lower voltage dropout as compared to a source follower pass transistor. However, it also has a lower phase margin, which can lead to ringing issues. The balance

4.4 Low-Dropout (LDO) Voltage Regulator

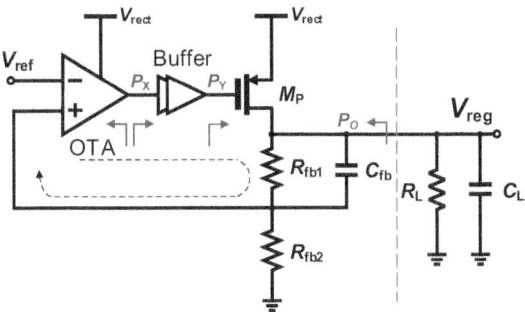

Fig. 4.10 Block diagram of a voltage regulator

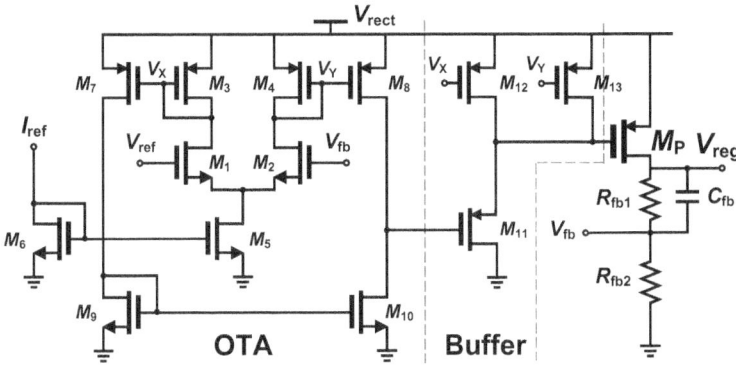

Fig. 4.11 Schematic of 1.8 V LDO (V_{reg})

between bandwidth and phase margin is determined based on the LDO output capacitance. The 1.8 V LDO comprises an EA (M_{1-10}), buffer (M_{11-13}), pass transistor (M_P, 3.3 V PMOS), and resistive feedback ($R_{fb1,2}$). The EA is designed based on an OTA structure with a bias current of 10 μA. To guarantee the operation of the PMOS pass transistor in the saturation region for maximum load current, the size of 3.2 mm/300 nm is chosen and in order to minimize the gate capacitance, the minimum channel length is used. Due to the large dimensions of the pass transistor, a large capacitor is connected to the output node of the EA, which significantly degrades the transient response of the LDO. To address this issue, a buffer is designed, which offers low output resistance.

A pole and zero analysis is performed to assess the stability of the regulator. The circuit has three important poles: at the output node of the EA (P_X), the gate of M_P (P_Y), and the output pole (P_O). Moreover, one zero is designed at the output (Z_{out}) [19].

$$P_X = \frac{1}{R_{OTA} C_X}, \quad P_Y = \frac{1}{(1/g_{m,buff}) C_Y}, \quad P_O = \frac{1}{R_O C_O} \quad (4.1)$$

where R_{OTA}, $g_{m,buff}$, C_X and C_Y, R_O, and C_O are the EA output resistor, the transconductance of the buffer, the capacitance at nodes X and Y, and the output resistor and capacitor at node O, respectively. The proposed LDO has a loop gain of 52.5 dB and a phase margin of 89° at 70 kHz when loaded with a 0.8 µF capacitor. The line regulation is the ratio between regulated voltage and supply voltage variations. Also, the load regulation is the LDO voltage change divided to a change in the loading current. The 1.8 V LDO exhibits 3.07 mV/V and 0.11 mV/mA line and load regulation, respectively, with a static power consumption of 409 µW.

4.4.2 1.2 V LDO with Capacitorless Output

External large capacitors are required at the output of LDOs to ensure stability and minimize voltage fluctuations. As power and size reduction trends drive system-on-chip integration, compatibility with CMOS technology is crucial for cost-effective solutions. This leads to research into CMOS capacitor-less low-dropout regulators and on-chip compensation techniques to maintain system stability and regulation performance. Portable devices prioritize low power consumption, but reducing static current can limit dynamic performance. To address this issue, transient enhancement circuit techniques are introduced to balance dynamic performance while minimizing the impact on power efficiency and circuit complexity.

An output-capacitor-less LDO has been designed in Fig. 4.12 to provide a 1.2 V supply voltage for the low-power blocks, which eliminates the need for a large output capacitor and minimizes power dissipation. The design focuses on low power consumption and minimum area. It comprises a folded-cascode error amplifier (FC-EA) optimized for low-voltage operations, with a PMOS input pair employed to maximize the linearity of the EA. The number of required bias voltages in the EA branches is reduced by using a self-bias method, where

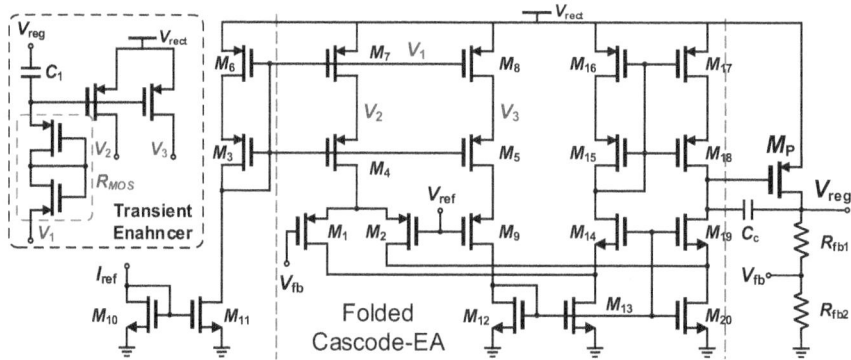

Fig. 4.12 Schematic of the 1.2 V output-capacitor-less LDO

4.4 Low-Dropout (LDO) Voltage Regulator

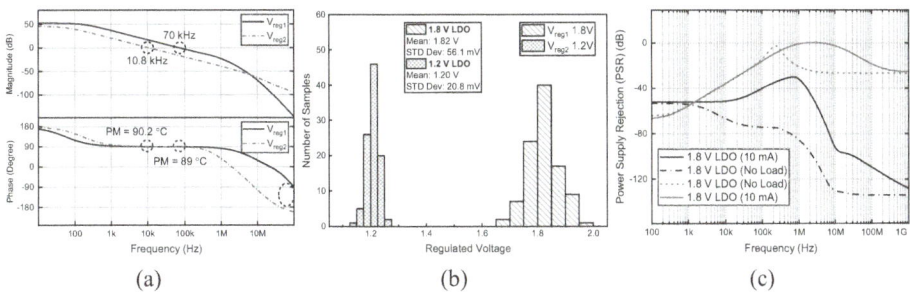

Fig. 4.13 **a** Simulated gain and phase frequency responses of the proposed LDOs, **b** Monte Carlo simulation results, and simulated power supply rejection (PSR)

the gates in the cascode branch are connected. To improve the transient response, the design activates dynamic currents in parallel to the currents at nodes V_2 and V_3 when the regulated voltage decreases, which accelerates the discharge process [20]. Additionally, a Miller compensation capacitance of 6 pF has been added to enhance the phase margin. The designed LDO has a loop gain of 45.6 dB and a phase margin of 90.2° at 10.8 kHz when loaded with a 100 pF on-chip capacitor, as shown in Fig. 4.13a. The line and load regulation of the LDO are 0.485 mV/V and 0.03 mV/mA, respectively, with a static power consumption of 875 μW. The Monte Carlo simulation results and power supply rejection of both LDOs are shown in Fig. 4.13b, c, respectively.

4.4.3 Measurement Results

The presented LDOs are fabricated in a 180 nm standard CMOS technology as shown in Fig. 4.14. Figure 4.15a displays the measured waveforms that emphasize the performance of the regulated and reference voltages, as well as their corresponding current dissipation

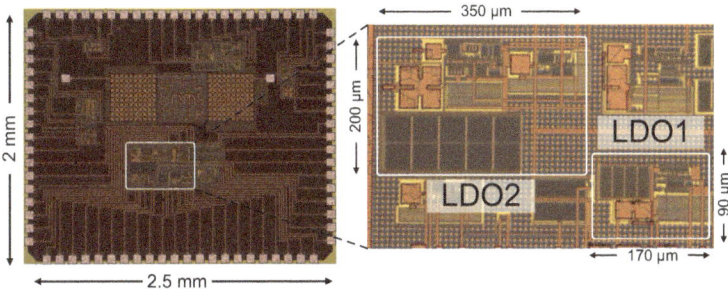

Fig. 4.14 Die microphotograph of the presented LDOs using a standard 180 nm CMOS process

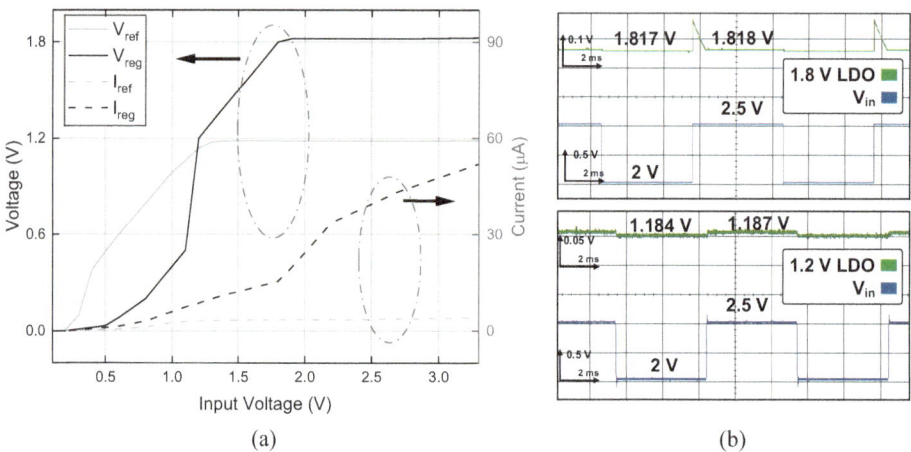

Fig. 4.15 a Measurement results of the regulated and reference voltages and their related current dissipation through sweeping the input DC voltage and **b** measured transient line regulation

at different input DC voltages. Additionally, Fig. 4.15b shows the measured transient line regulation.

4.5 Power Control Unit

Safety is a critical concern in wirelessly powered implants, with biomedical standards regulating WPT through the human body. This includes assessing electromagnetic field absorption and temperature elevation effects [21, 22]. Additionally, power regulation is essential to prevent excessive dissipation and ensure compatibility with implant electronics, as shown in Fig. 4.16. Remote powering becomes unacceptable in two cases: excessive power delivery, leading to wasted energy, temperature elevation, and potential circuit damage; and insufficient power, causing system failure. Various closed-loop WPT solutions address these issues. In [23], a single inductive link with feedback adapts power transmission to implant needs. [24] proposes self-regulating WPT with automatic amplitude control, balancing transient response and voltage accuracy. Self-regulating and reconfigurable resonant rectifiers [25, 26] achieve regulation via capacitor charging cycles but suffer from low efficiency and coupling range limitations. Other closed-loop feedback designs, such as [27], use peak detectors and differential amplifiers for voltage control. Overvoltage protection techniques, including input signal limiting, secondary terminal shortening, and LC tank detuning, are explored in [28], but may disrupt power stability. Protection circuits [28, 29] and implant battery voltage/current control mechanisms [30] further enhance power management.

4.5 Power Control Unit

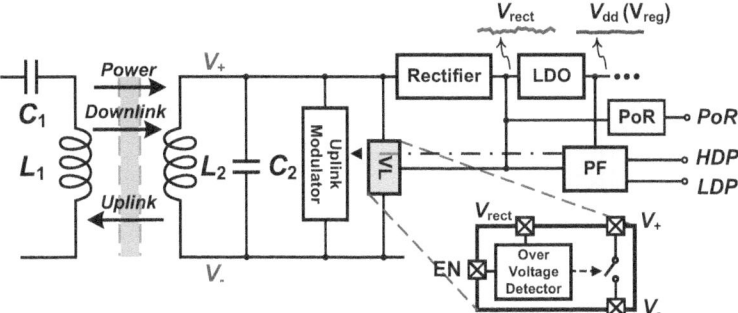

Fig. 4.16 Simplified block diagram of the presented power control unit

4.5.1 Power Feedback (PF)

To ensure safe and efficient WPT, a power feedback control unit plays a critical role. This unit facilitates closed-loop monitoring, preventing excessive or insufficient power delivery, as illustrated in Fig. 4.17. By continuously monitoring the rectified voltage, the system identifies inefficiencies in wireless power transfer and dynamically adjusts the transmitted power. Furthermore, the acquired data can be relayed through an uplink communication path to an external unit or the implant's controller.

The feedback mechanism operates by comparing a divided regulated voltage—a stable and noise-free DC reference—with a voltage derived from rectification and division via two distinct OTAs. On-chip resistors perform the voltage division. The control unit evaluates the rectifier's output voltage and indicates power delivery status using two output bits: high delivered power (HDP) and low delivered power (LDP). Specifically, the upper comparator asserts the HDP bit when V_C exceeds V_A, while the LDP bit is activated when V_C falls below V_B. Additionally, a current reference generator is integrated to provide appropriate bias currents for the OTAs. Threshold values for HDP and LDP detection are analytically

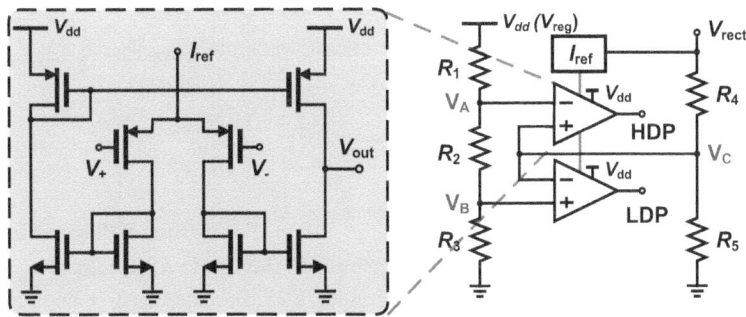

Fig. 4.17 Schematic of the power feedback (PF)

determined, assuming a rectified voltage range of 2–2.2 V. The HDP and LDP bits are transmitted via an uplink communication path, allowing real-time adjustment of inductive link drivers, such as power amplifiers. To optimize power efficiency, the resistor values (R_{1-5}) are carefully selected to minimize energy consumption.

According to Fig. 4.17, the voltage relations are derived as follows:

$$V_A = V_{reg} \frac{R_2 + R_3}{R_1 + R_2 + R_3}$$
$$V_B = V_{reg} \frac{R_3}{R_1 + R_2 + R_3}, \quad \begin{cases} V_C \geq V_A \Rightarrow \text{HDP} = 1 \\ V_C \leq V_B \Rightarrow \text{LDP} = 1 \end{cases} \quad (4.2)$$
$$V_C = V_{rect} \frac{R_5}{R_4 + R_5}$$

$$\Rightarrow \begin{cases} V_{rect} \geq V_{reg} \frac{(R_4+R_5)(R_2+R_3)}{R_5(R_1+R_2+R_3)} & \text{HDP} = 1 \\ V_{rect} \leq V_{reg} \frac{(R_4+R_5)R_3}{R_5(R_1+R_2+R_3)} & \text{LDP} = 1 \end{cases} \quad (4.3)$$

The resistor values can be determined based on an assumed regulated voltage of $V_{reg} = 1.8$ V, with HDP and LDP threshold limits set at 2.45 V and 2 V, respectively. For the implantable medical device (IMD), two scenarios may arise when the delivered power is insufficient (LDP = 1). In the first scenario, despite inadequate received power, the regulated voltage remains stable due to the presence of an internal energy source and the line and load regulation characteristics of 3 mV/V and 0.11 mV/mA, respectively. Consequently, no additional considerations are required. In the second scenario, the regulated voltage is not maintained because the rectified voltage fails to meet the regulation criteria. As a result, the regulated voltage directly follows the rectified voltage ($V_{rect} \approx V_{reg}$). Under this condition, the LDP state, as defined in (4.2), necessitates an additional condition:

$$1 \leq \frac{(R_4 + R_5)R_3}{R_5(R_1 + R_2 + R_3)} \quad (4.4)$$

Figure 4.18 shows a simplified model of the PF unit. The system can be analyzed by considering the model of the rectifier and LDO. Assuming that $X_{\text{HDP}} = \frac{(R_4+R_5)(R_2+R_3)}{R_5(R_1+R_2+R_3)}$ and $X_{\text{LDP}} = \frac{(R_4+R_5)R_3}{R_5(R_1+R_2+R_3)}$, when the regulation condition is met ($V_{rect} \geq 1.8$ V):

$$\Rightarrow \begin{cases} V_{rect} \geq V_{reg} \frac{X_{\text{HDP}}}{1-X_{\text{HDP}}\alpha V_{reg}} & \text{HDP} = 1 \\ V_{rect} \leq V_{reg} \frac{X_{\text{LDP}}}{1-X_{\text{LDP}}\alpha V_{reg}} & \text{LDP} = 1 \end{cases} \quad (4.5)$$

where α is the line regulation of the regulator. When the rectified voltage is between 1.3 and 1.8 V, the pass transistor of the LDO operates in the triode region and $V_{reg} = (V_{rect} - V_{\text{LDO}}) \approx V_{rect}$, where V_{LDO} is the dropout voltage at the LDO and is negligible in this operation region.

4.5 Power Control Unit

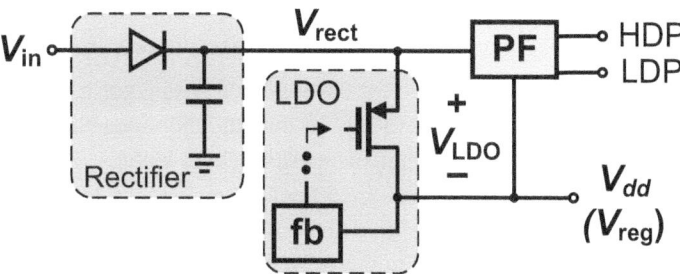

Fig. 4.18 Simplified model of power feedback unit

4.5.2 Power on Reset (PoR)

The PoR block initiates data communication when it detects that the delivered power is sufficient for robust communication. This is necessary because in the initial phase of wireless power transmission, the power level starts at zero and requires a certain amount of time to reach a sufficient level by charging a capacitor. However, other circuits within the system also consume current and prolongate the settling time [31]. The PoR circuit temporarily disables these circuits, such as the downlink demodulation circuit, until the capacitor voltage reaches a predefined level. If the output of the LDO falls below a designated threshold, the PoR deactivates other circuits to conserve charge on the capacitor and minimize unnecessary current consumption. The schematic of the PoR is presented in Fig. 4.19a. As V_{rect} increases, transistor M_2 is triggered via the gate of M_1. Through positive feedback, the third-stage transistor is turned off, forcing the inverter input to a high state, which drives PoR to GND and enables all data conversion circuits to operate. The circuit functions with a power consumption of 72 μW and deactivates the PoR mechanism once V_{rect} reaches 1.75 V.

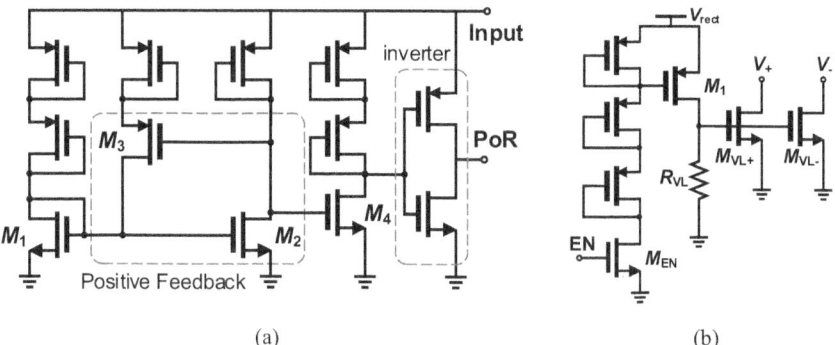

Fig. 4.19 Schematic of the **a** power-on-reset (PoR) and **b** voltage limiter (VL)

4.5.3 Voltage Limiter (VL)

At the implant site, the rectified voltage can exceed 3.3 V, posing a risk of electronic circuit damage and potential temperature rise [32]. To mitigate this overvoltage issue, a VL or overvoltage protection circuit can be implemented using high-voltage (3.3 V) IO transistors, as depicted in Fig. 4.19b. The VL circuit incorporates diode-connected MOSFETs, which turn on when the voltage surpasses a predefined threshold. This activation establishes a parallel current flow path in addition to the inductive link, effectively dissipating excess energy and preventing excessive power accumulation. The circuit employs three diode-connected transistors to minimize the current dissipated in that branch and protect each transistor from excessive voltage. The dimensions of the transistors and the value of the resistor (R_{VL}) are calculated to minimize the loading on the rectifier. An additional enable switch (NMOS) is placed inside the circuit to control when VL should be activated; this switch is controlled by the implant controller and power feedback unit. This transistor operates in the triode region ($V_{DS,M_{VL8}} \simeq 0$). The circuit can effectively protect the electronic circuitry from overvoltage with minimum static power dissipation.

4.5.4 Measurement Results

The control units are designed and fabricated in a 180 nm standard CMOS technology, as shown in Fig. 4.20. Initially, a triangular waveform with an amplitude range of 1.3–2.7 V is applied to the input of the power feedback unit and the LDO (representing the rectified voltage) to evaluate the HDP and LDP output responses along with the regulated voltage. The experimental results are presented in Fig. 4.21a. The measured response times for the HDP and LDP bits during rising and falling transitions are 18.6 μs and 21.4 μs, respectively, indicating that the feedback system identifies unacceptable power transmission within approximately 20 μs. Furthermore, the input DC voltage is swept to analyze system

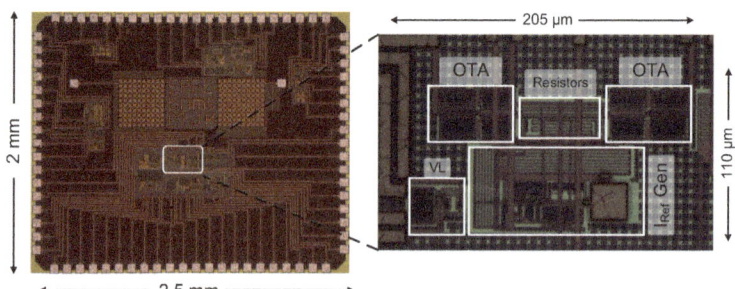

Fig. 4.20 Die micrograph of the proposed power feedback unit in a standard 180 nm CMOS process, occupying an area of 205 μm × 110 μm

4.5 Power Control Unit

behavior and determine the threshold voltages for the HDP and LDP bits, as shown in Fig. 4.21b, along with the corresponding simulation results. The HDP bit is asserted when the input voltage exceeds 2.45 V, whereas the LDP bit is activated when the input voltage falls below 2 V. The measurement results closely align with the simulation data, validating the circuit's expected performance. Additionally, the power dissipation of the feedback unit is evaluated, as illustrated in Fig. 4.21c. Within the acceptable input voltage range, the circuit

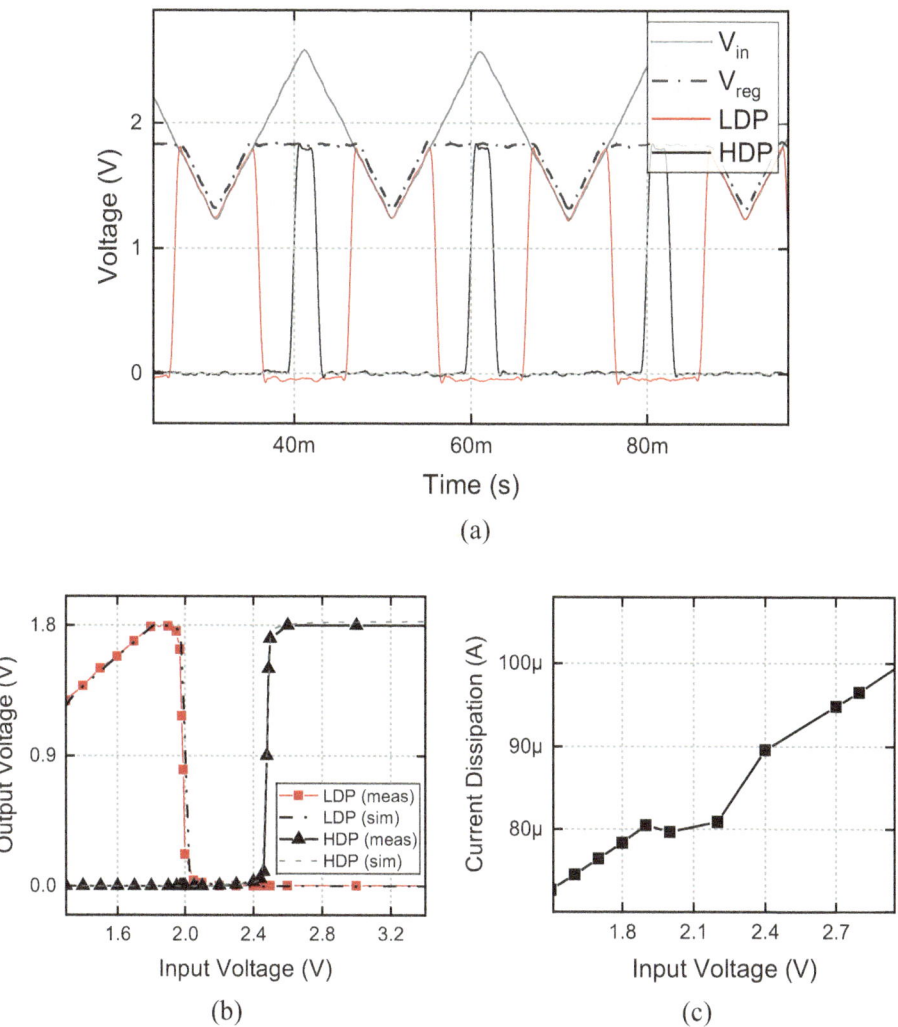

Fig. 4.21 Measured waveform of **a** the input and output voltages of the power feedback unit, **b** measured and simulated results of the input voltage sweep for the power feedback unit, and **c** measured current dissipation in the power feedback unit as a function of input DC voltage sweep

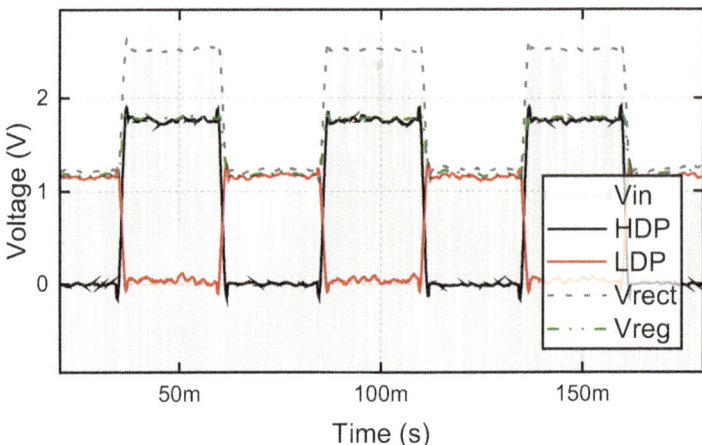

Fig. 4.22 Measured results of the transient response of the power feedback unit under wireless powering conditions

consumes 80 μA, resulting in a power dissipation of 144 μW. The power feedback system is further assessed under wireless conditions at 13.56 MHz to examine its impact on power conversion efficiency. The measurement outcomes, displayed in Fig. 4.22, include rectified and regulated voltage values alongside HDP and LDP control signals. When the feedback unit is connected, the voltage rectifier output decreases by 3.3% in the experimental setup, whereas the simulation predicts a reduction of 0.2%. The system operates correctly with a minimum required input voltage of 1.3 V.

Table 4.2 summarizes the key performance of power feedback units in the wirelessly powered implants reported in the literature. Moreover, it demonstrates the diverse possibilities for wireless power regulation on either the transmission (Tx) or receiver (Rx) side, depending on the specific application. While regulating rectifiers serves as an efficient method for power monitoring and regulation, they possess limitations in preventing excessive power transmission or alerting the primary side to this concern. To effectively demonstrate the performance of the voltage limiter, a 10 kΩ resistor (connected to V_{DD}) is placed in series with V_+ to facilitate current measurement. The simulated and measured resistor current, obtained by performing a DC sweep on the rectified voltage, along with the corresponding setup schematic, is presented in Fig. 4.23. The voltage limiter is activated when the rectifier output voltage exceeds 3.53 V. Prior to activation, its power consumption remains minimal at 0.7 μW, ensuring negligible impact on the system's overall efficiency.

4.6 Wireless Power Transfer Unit

Table 4.2 Power feedback unit performance comparison [33]

	Tech (μm)	f (MHz)	Method[a]	VR[b] (V)	#Bits[c]	Site[d]
[23] JSEN'13	0.18	8	Bits	1.58 | 1.62	2	Tx
[25] JSSC'17	0.18	144	R^2	0.8 | 1.6	1	Rx
[34] JSSC'17	0.35	6.78	R^2	–	3	Rx
[35] TBCAS'19	0.18	144	R^2	−0.4 | 0.4	4	Rx
[36] ISCAS'21	0.18	433	SR	1.7 | 1.9	2	Rx
[37] JSSC'21	0.25	6.78	R^2	–	4	Tx/Rx
[26] TPE'23	0.18	6.78	R^2	–	2	Tx/Rx
This work	0.18	13.56	Bits	2 | 2.45	2	Tx/Rx

[a] Method, bits: digital control bits, R^2: regulating rectifier, SR: self-regulation
[b] Controlled output voltage range
[c] Number of control bits
[d] Power management and control site

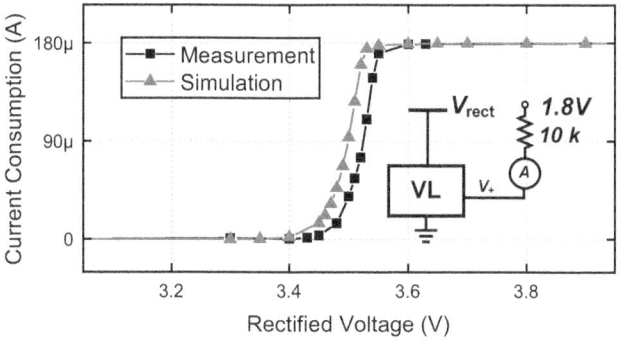

Fig. 4.23 Measurement and simulation results of the current dissipation in the voltage limiter (VL) as a function of the input rectified voltage sweep

4.6 Wireless Power Transfer Unit

A wireless power transmitter consists of two stages. The first is an oscillator that generates the required pulse signal for power transmission in the inductive links. The second is a power amplifier that delivers AC current into the transmitter coil for wireless power transmission. For generating stable frequency signals for clocking and data transmission, an oscillator

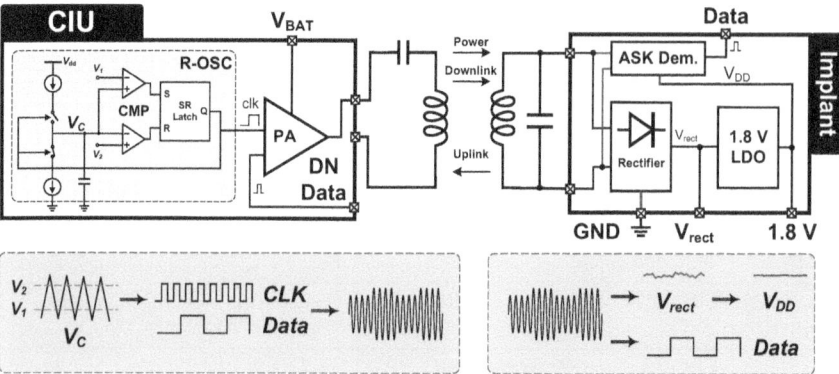

Fig. 4.24 Block diagram of the proposed wireless power transfer in a central implanted unit (CIU) including the inductive link, relaxing oscillator (R-OSC), and power amplifier (PA)

is required. Relaxation oscillators (R-OSCs) have efficient power consumption and small area characteristics with the ability to precisely set their oscillation frequency [38–40]. The block diagram of the WPT unit, which operates using a single pair of coils at 13.56 MHz, is shown in Fig. 4.24. The system includes an R-OSC generating the required clock and a PA to transfer power and modulated data. Also, the WPT unit in the CIU is supplied by an external battery. A buck DC-DC converter is designed to step down the battery voltage for other electronic circuits.

4.6.1 Power Amplifier (PA)

A PA is designed in this WPT system to transfer from the CIU to the patches through the inductive link. A class-E PA is chosen for this purpose due to its high efficiency in driving the inductive link, if it is nominally tuned [41, 42]. As shown in Fig. 4.25, the class-E PA is connected to the inductive link. Transistor M_1 acts as a switch, periodically activated by a clock signal at the gate voltage. The RF choke L_{RFC} has a high impedance at the operational frequency, serving as a DC current source. While M_1 is in the ON state, L_{RFC} is positioned between V_{DD} and ground, accumulating energy from V_{DD}. When M_1 is off, L_{RFC} transfers the accumulated energy from V_{DD} into the load network [43]. The value of L_{RFC} (12 μH) is selected to be larger than L_{Tx}. The values of C_P (188 pF) and C_{Tx} are calculated to achieve nominal tuning for the PA. In a nominally-tuned setup, when M_1 is activated, V_D decreases to zero. This condition indicates that the PA meets zero voltage switching (ZVS) criteria. Additionally, the schematic of a class-E PA with an ASK modulator is shown in Fig. 4.25. The carrier signal connected to M_{PA1} provides the operation frequency of the wireless power transmission. Also, M_{PA2} and M_{PA3} switches are employed for the ASK data modulation.

4.6 Wireless Power Transfer Unit

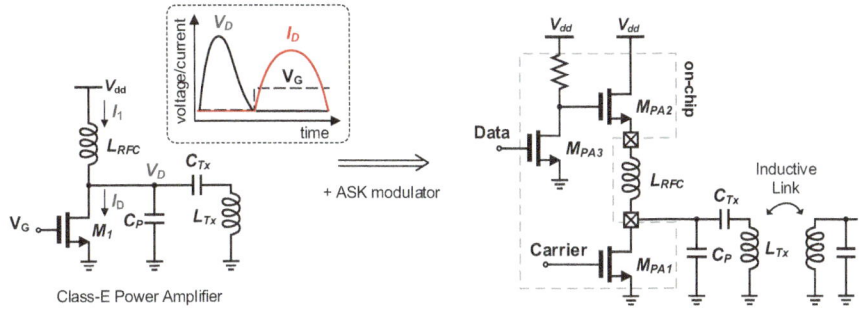

Fig. 4.25 Schematic of the class-E power amplifier (PA)

4.6.2 Relaxation Oscillator

R-OSCs can be characterized by the inclusion of a comparator featuring hysteresis. These R-OSCs are also known as RC oscillators. The fundamental operation of this oscillator relies on the comparator's input transitioning between two hysteresis levels, which receive feedback from the comparator's output. The feedback mechanism determines the input's direction, and the comparator's output switches when the input reaches these specified levels. The capacitor located before the input of the comparator can be charged or discharged using either an RC circuit or a charge pump circuit, offering flexibility in the design. The circuit diagram of the proposed R-OSC is shown in Fig. 4.26. This R-OSC operates by either charging or discharging a capacitor with a constant current source. Charging the capacitor with a constant current source results in a constant slope, leading to the generation of a

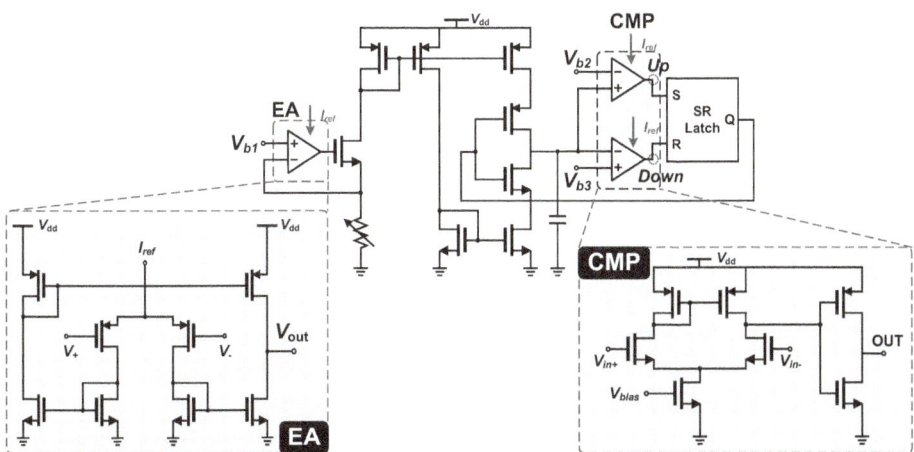

Fig. 4.26 Schematic of the relaxation oscillator implemented in a CIU

triangle wave at the comparator's input. The design incorporates two comparators and an SR latch, which collectively function as a hysteresis comparator. The voltage levels, denoted as V_0 and V_1, serve as the hysteresis thresholds that trigger output toggles. When the triangle wave reaches these voltage levels, a brief single pulse is generated to set or reset the output via the SR latch. The SR latch maintains the output value until another pulse modifies it. The period at which nodes get back to their initial state has to be calculated to determine the oscillation frequency. Since the charging and discharging currents are equal, the duty cycle remains at 50%. Considering the oscillator's intended frequency range of tens of MHz, the charging time constant is set to one-thousandth of the oscillation period to ensure linear charging. Consequently, the on-resistance needs to be in the order of a few hundred Ohms.

The oscillator is designed to operate within the frequency range of 5–20 MHz, with target oscillation frequencies of 13.56 and 6.78 MHz. This frequency accounts for variations in the manufacturing process, necessitating the incorporation of a calibration loop. The voltage across the capacitor (V_{cap}) is compared to two hysteresis voltages, with the comparators generating set and reset signals labeled as "*Up*" and "*Down*." These signals are then linked to the SR latch, which also serves as the output of the design. This oscillator exhibits an average power consumption of approximately 110 µA for a 13 MHz oscillation frequency, resulting in a power consumption of 198 µW when supplied with a voltage of 1.8 V.

4.6.3 DC-DC Converter

Battery-powered electronic devices require the conversion of the battery voltage to the desired levels with minimal energy loss. This is particularly important as the battery voltage remains relatively constant. Hence, step-down voltage conversion is essential in most applications. Switching mode DC-DC buck converters have the potential to achieve higher efficiency compared to linear regulators [44]. The power stage of this converter, as shown in Fig. 4.27, consists of NMOS and PMOS transistors, which are synchronously controlled. Consequently, when one transistor is turned on, the other is turned off. The purpose is to convert the battery voltage into a stable 1.8 V supply for the electronic components within the CIU.

Figure 4.27a shows the block diagram of a voltage-mode PWM buck converter. The inductor and capacitor are used to construct a filter and the output signal V_{PWM} from the comparator is responsible for generating both gate voltages of switches, which control the activation and deactivation of the high-side PMOS transistor and low-side NMOS transistor, respectively. When the high-side transistor is conducting, the current flowing through it is equal to the inductor current. This current increases at a positive slope determined by the inductor value and output voltage signal. When the ramp generator signal exceeds the control voltage (V_{ref}), the comparator output signal goes high. Similarly, the PWM signal causes both V_P and V_N to go high, turning off the high-side PMOS transistor and turning on the low-side NMOS transistor until the next clock pulse is received through the Non-overlap and

4.6 Wireless Power Transfer Unit

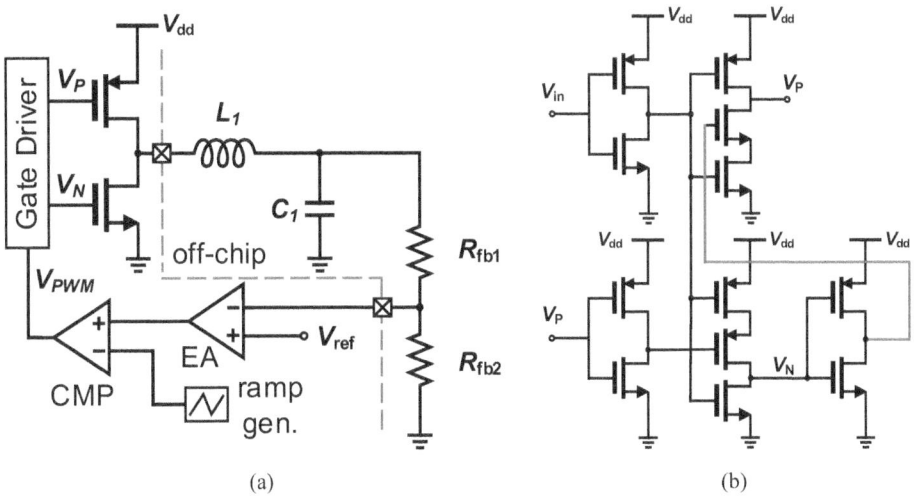

Fig. 4.27 Schematic of the DC-DC converter in a CIU

driver circuit. In this design, the employed ramp generator is a modified version of the one presented in Fig. 4.26. Additionally, the design utilizes the 1.2 V reference voltage shown in Fig. 4.7. The driver buffer stage shown in Fig. 4.27b is responsible for delivering the controlling pulses to the power transistors [44]. The input signal for the buffer is obtained from the output of the comparator (V_{PWM}). If a basic CMOS inverter is employed as a buffer, there is a possibility of short circuits and energy losses occurring when both the NMOS and PMOS power transistors are conducting simultaneously. However, the buffer topology used in this case prevents these short-circuit losses by introducing a short gap time where both the NMOS and PMOS transistors are switched off. Figure 4.28a shows the transient simulation results of the designed DC-DC converter, including the input voltage (battery), reference voltage (1.2 V), output voltage (1.8 V), ramp, and control voltage. The design achieves output voltage stabilization within a 4 μs timeframe. Figure 4.28b displays the simulated power conversion chain of the DC-DC converter through various loads.

4.6.4 Measurement Results

The wireless power transfer unit is designed and fabricated in a 180 nm standard CMOS technology, as shown in Fig. 4.29. Figure 4.30a shows the measurement results of the power and data transmission using the R-OSC clock feeding the designed class-E PA, with the spectrum of the clock and 40 kilobits per second (kbps) downlink data communication. Subsequently, this voltage waveform serves as the energy source for the inductive link implemented on the

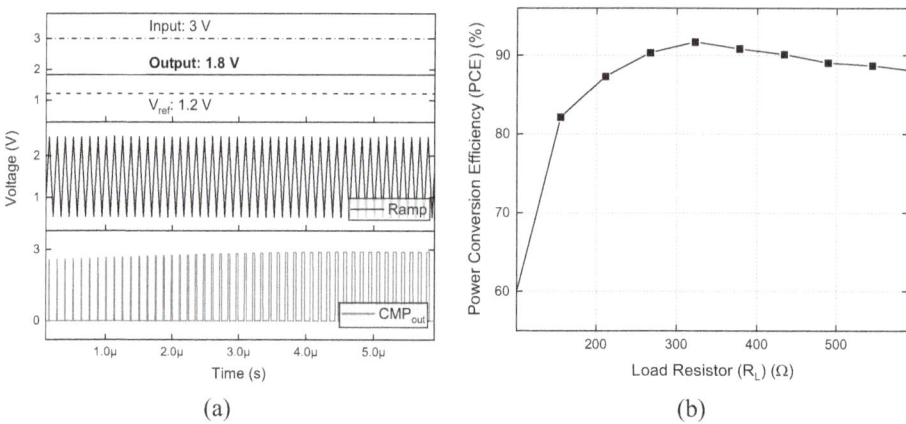

Fig. 4.28 Simulation result of the **a** transient analysis of the DC-DC converter **b** power conversion chain through various loads

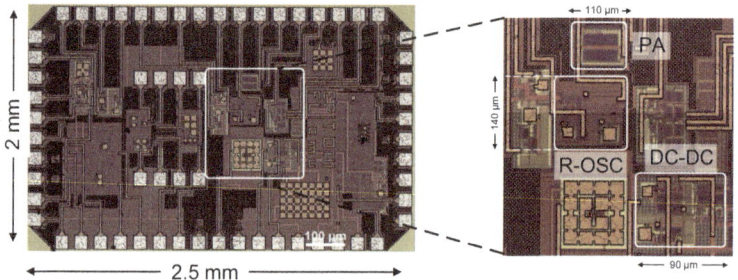

Fig. 4.29 Die microphotograph of the presented wireless power transfer unit using a standard 180 nm CMOS process

PCB to transmit power and ASK-modulated data to the implant. Figure 4.30b shows the 1.8 converted voltage from the battery voltage (3.2 V).

4.7 Automatic Resonance Tuning System

Achieving a sufficiently high PTE is essential to minimize energy dissipation in biological tissue for safety considerations. Maximizing PTE is typically accomplished by estimating the load of the secondary unit and matching the quality factors of both the primary and secondary stages [45]. Several studies have explored theoretically optimal circuit configurations to enhance PTE. For instance, [45] analyzes optimal resistive loading conditions, while [46] proposes an optimal resonant load transformation approach applicable to resonators with fixed loading impedance. In this framework, the resonance capacitor is connected in series

4.7 Automatic Resonance Tuning System

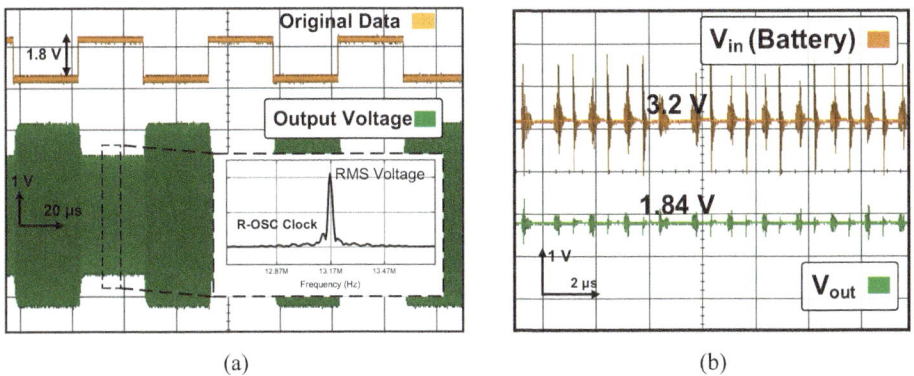

Fig. 4.30 Measurement results of **a** wireless power transfer unit and **b** DC-DC converter

with the inductor, and additional components are introduced in parallel with the resonator to optimize performance [33]. The values of these components are derived based on the optimal quality factors. Despite meticulous design, practical variations in factors such as distance, temperature, and environmental conditions are inevitable. These discrepancies can shift the resonant frequency from its intended value and lead to mismatches between the primary and secondary resonance frequencies, thereby reducing PTE. The mismatches encountered in WPT systems can generally be classified into three categories: (a) LC variations, where the inductance remains constant, and variations occur in the capacitance; (b) distance fluctuations, affecting the coupling coefficient between coils; and (c) angular misalignments, altering the mutual inductance between the transmitter and receiver. While prior research has primarily focused on mitigating LC variations, limited attention has been given to evaluating the effectiveness of tuning systems in compensating for distance variations and angular misalignments. The following sections provide a summary of existing tuning mechanisms implemented on either the primary or secondary side, most of which rely on tunable capacitors to adapt to changing conditions.

Automatic tuning systems can be categorized based on their tuning methodology-such as successive approximation register (SAR) logic [47, 48] and monotonic sweeping [49–53]-as well as their detection target, which may involve monitoring the rectified voltage/current or swing amplitude. Monotonic sweeping utilizes a counter and control logic to dynamically adjust the resonance capacitor, whereas SAR logic employs a binary search algorithm for automatic tuning. In battery-powered systems, rectified current can serve as the detection target, optimizing efficiency through the integration of a current sensor [53]. Alternatively, instead of relying on rectified voltage measurements, swing amplitude monitoring offers the advantage of reducing delays associated with voltage stabilization and rectifier output capacitor charging. The design of automatic tuning systems is more extensively explored on the primary side, as external units are not implanted and can accommodate additional functional blocks without strict constraints on size and power dissipation. Early implementations primarily involved air-coupled inductive links, some of which required direct communication between the primary and secondary units.

Since implanted devices impose constraints on microcontroller size and power consumption for resonance tuning, many designs favor dedicated custom circuits tailored specifically for tuning operations. In [54], a real-time capacitor compensation scheme is introduced, where the phase shift in the secondary coil's current is detected via a dual-edge sampler, and the resonant coupling capacitor is dynamically adjusted using binary-weighted capacitor switching. Although this method is claimed to be faster than binary search, as it limits capacitance variation to 25%, the power and clock sources for the secondary coil appear to be externally supplied, making the implementation impractical for real-world implantable medical devices (IMDs). In [48], SAR logic is employed to control capacitor switching, complemented by start-up circuits. Rather than utilizing binary-weighted capacitors, this design incorporates multiple switched capacitor units connected to higher bits of the SAR logic, ensuring a more uniform layout and facilitating parasitic capacitance monitoring. It is inferred that the two clock signals serve a dual purpose-driving the SAR logic and controlling the switching mechanisms of the variation sensor described in [48]. Additionally, recent studies have explored active compensation techniques, such as those discussed in [55–58], where capacitor banks are adaptively tuned to counteract resonance variations.

This section presents a 13.56 MHz automatic resonance tuning system designed for wirelessly powered biomedical implants, as depicted in Fig. 4.31. One of the primary challenges faced by wirelessly powered IMDs-resonance variations in the secondary coil-is effectively addressed through a counter-based automatic resonance tuning technique, which offers a compensation dynamic range of six control bits. The system is capable of tolerating mismatches equivalent to a 75 pF capacitance variation in the LC tank while consuming only 154.7 μW of power during resonance tuning. Furthermore, the system's performance has been validated in handling variations in both coil distance and angular misalignment.

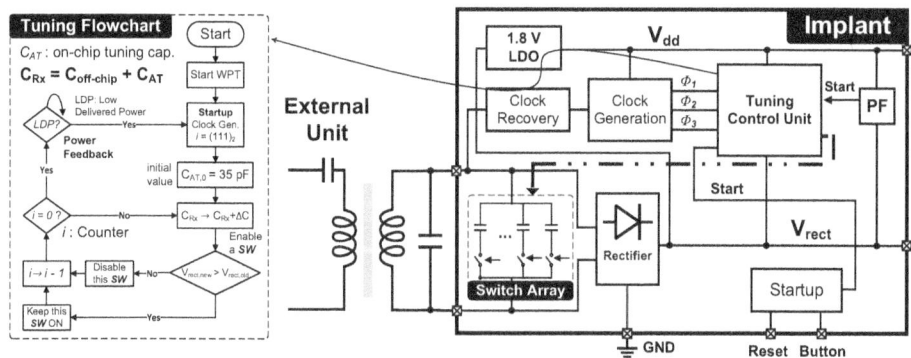

Fig. 4.31 Block diagram of the presented wirelessly powered resonance tuning system along with its tuning flowchart

4.7.1 Properties of Inductive Links

The schematic representation of a simplified inductive link is illustrated in Fig. 4.32. The design of the coil plays a crucial role in determining the coupling coefficient, which directly impacts the maximum voltage gain between resonators in a series-parallel topology. When the coupling coefficient exceeds a critical threshold, the voltage gain decreases, potentially leading to instability due to energy being transferred back to the primary coil. It is important to note that in practical biomedical applications, achieving strong coupling is not always feasible, as variations in implant positioning and tissue properties can affect the inductive link performance. Derivations of the voltage gain $|v_{out}/v_{in}|$ have been studied in the frequency domain in [45] and the transfer function is expressed as

$$\frac{v_{out}}{v_{in}} = \left(\frac{(1/k)L(j\omega)}{1-L(j\omega)}\right)\left(\frac{1}{j\omega\sqrt{L_1 L_2}}\right)\left(\frac{R_L}{j\omega C_2 R_L + 1}\right) \quad (4.6)$$

where ω, $L_{1,2}$, $C_{1,2}$, k, and R_L are the carrier frequency, primary and secondary inductors and capacitor values, coupling coefficient, and load resistor. The loop transmission $L(j\omega)$ is

$$L(j\omega) = \frac{(j\omega)^2 k^2 L_1 L_2}{Z_1 Z_2} \quad (4.7)$$

The impedance of the primary and secondary resonators is simplified as

$$Z_{1,2} = \frac{(j\omega)^2 + j\omega \frac{\omega_c}{Q_{1,2}} + \omega_c^2}{j\omega / L_{1,2}} \quad (4.8)$$

where ω_c is the resonance frequency and $Q_{1,2}$ is the quality factor of the inductors.

Although the formula for maximum PTE has been derived in [45] under the assumption of matched load resistance, it is not directly utilized in this system due to the fixed load imposed by the rectifier. The theoretical voltage gain at resonance as a function of the coupling factor k is plotted in Fig. 4.33a. A similar decline in voltage gain at strong coupling, as illustrated in Fig. 4.33b, confirms the consistency between theoretical predictions and simulation results. As previously discussed, critical coupling refers to the coupling condition at which voltage

Fig. 4.32 Schematic of a series-parallel (SP) inductive link

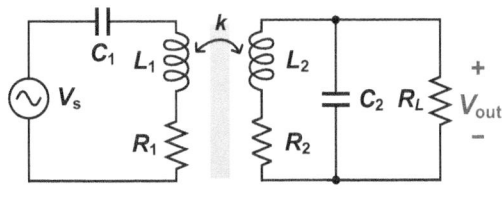

External Unit   **skin**   Implant

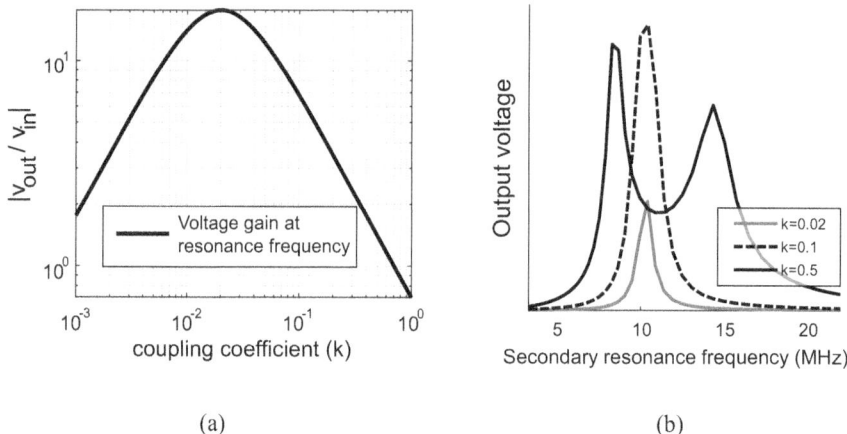

Fig. 4.33 a Calculated voltage gain at the resonance frequency with sweeping coupling factor; **b** qualitative simulation results of the output voltage with sweeping the resonance frequency

gain is maximized, with the corresponding coupling factor denoted as k_c. When the coupling factor satisfies $k < k_c$, frequency sweeping simulations exhibit a distinct peak at resonance. However, for $k > k_c$, a saddle-shaped response emerges, characterized by reduced voltage gain at resonance, as demonstrated in Fig. 4.33b. This separation of peak voltage gains is known as the frequency splitting phenomenon, which has been exploited in certain studies for communication purposes despite its inherent reduction in PTE [59].

Given that the automatic tuning system in this section aims to maximize PTE, the operating frequency is maintained close to the resonance frequencies of both the primary and secondary coils. To compensate for potential mismatches between these frequencies, deviations are modeled by sweeping the tuning capacitance C_2 (as shown in Fig. 4.32) around its optimal value. Figure 4.34 presents the resulting PTE, output voltage, and the derivative of the output voltage with respect to tuning capacitance, providing an evaluation of system performance under varying tuning conditions. The tuning strategy proposed in [48] leverages the derivative of the output voltage with respect to the resonance capacitance-where a positive derivative indicates the need to increase capacitance, and a negative derivative suggests the opposite. For cases involving a monotonically increasing derivative curve at high coupling levels, the coupling factor k is likely unrealistically large. In such scenarios, limiting the tuning time, as proposed in [48], can help mitigate the issue. A more detailed examination is conducted on the relationship between the peak PTE and the tuning capacitance at the points where the derivative crosses zero. As coupling weakens, the difference between the two PTE peaks diminishes. Since this work does not employ a reference voltage and primarily focuses on operation under loose coupling conditions, maximizing the output voltage serves as an effective approximation for optimizing PTE.

4.7 Automatic Resonance Tuning System

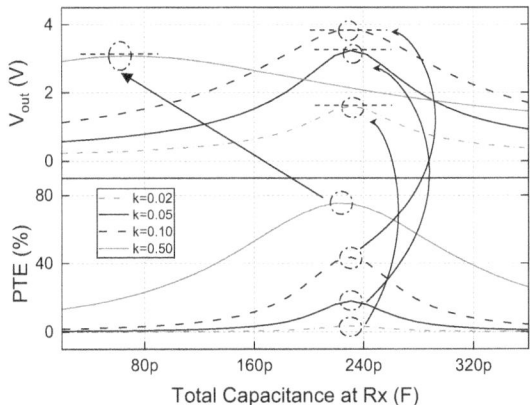

Fig. 4.34 Power transfer efficiency (PTE), output voltage and its derivative with respect to the tuning capacitance C_t

Table 4.3 Denotation of signals in the automatic resonance tuning system [60]

Denotation	Definition
V_+, V_-	AC signal coming from the inductive link
V_{rect}	Rectified voltage
V_{in+}, V_{in-}	Input voltages of the latched comparator used in the tuning control unit
V_{CMP+}	Output of the latched comparator, $V_{CMP+} = 1$ when $V_{in+} > V_{in-}$
Φ_1 to Φ_3	Clock with same period but different duty-cycles
$SW\langle 5:0\rangle$	Control signals to the switched capacitors
$C_{SW}\langle 5:0\rangle$	Six switched capacitors
$\overline{RST_{in}}$	Reset signal of the DFFs
C_{prev}	Storage capacitor of the rectified voltage

The flowchart of the auto-tuning algorithm is depicted in the right box of Fig. 4.31. This algorithm effectively stabilizes the rectifier voltage within a predefined range, even in the presence of external disturbances affecting the WPT link. A summary of signal notations is provided in Table 4.3. The schematic representation of the auto-tuning system is shown in Fig. 4.35, where the secondary resonance frequency is dynamically adjusted through the activation or deactivation of six switched capacitors. As illustrated in Fig. 4.35, the control unit consists of a comparator, a three-bit down counter, a one-hot decoder, six identical processing units, and clock recovery and clock divider circuits. The start-up and termination of the control unit are managed by the counter and an input signal $\overline{BTN}$, which acts as a pulse signal mimicking a button press to initiate the tuning process. The design and functionality of these modules are further detailed in the following sections.

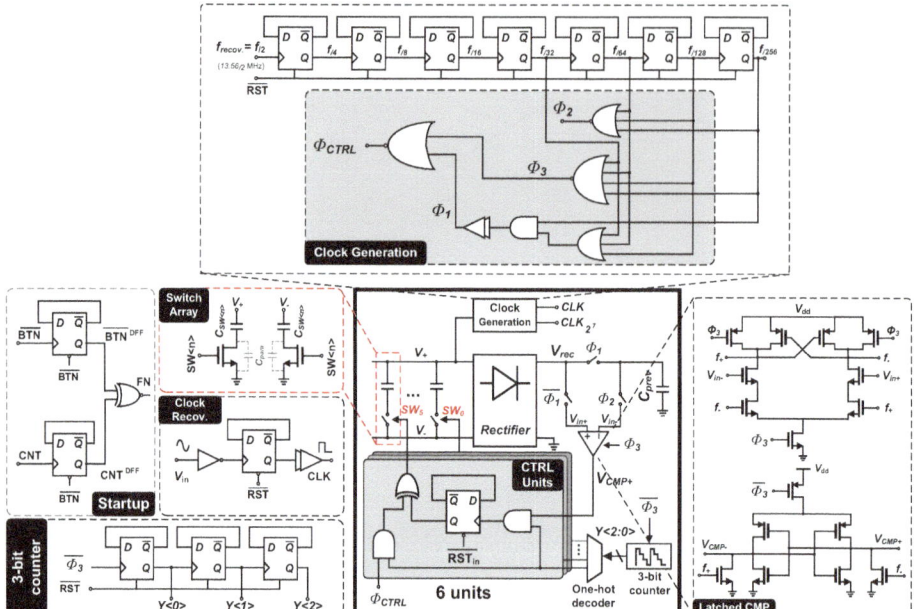

Fig. 4.35 Schematic of the designed automatic resonance tuning system, including the capacitor bank, logic control units, switch array, and rectifier

4.7.2 Capacitor Bank

Six switched capacitors are connected in parallel with the resonance capacitor, contributing 30% of the ideal resonance capacitance $C_{2,opt}$ at the secondary to compensate for a bidirectional 15% LC variation. For an implant coil with a 0.6 µH inductance, achieving resonance at 13.56 MHz requires a capacitor range of 69 pF. The system is designed with an overall capacitor coverage of 75 pF, ensuring sufficient margin. A trade-off exists between tuning time, accuracy, and the maximum achievable output voltage. To balance these factors, a tuning resolution of 5 pF is selected, optimizing both response time and energy consumption. Bidirectional tuning is implemented by inserting inverters before the switched capacitors at indices 5, 3, and 0, though not depicted in the diagram. A switched-capacitor structure is employed, with the parasitic capacitance of transistors highlighted in the red rectangle of Fig. 4.35.

4.7.3 Start-Up and Termination

The circuitry of the start-up and termination module is depicted in Fig. 4.35, specifically in the top left box, where a finished signal (FN) is incorporated. Several D flip-flops (DFFs)

are reset using $\overline{RST}$, which is defined as $\overline{RST} = \overline{RST_{in}} \cdot FN$. The waveform representation in Fig. 4.36a illustrates the operation of this module when the universal reset signal $\overline{RST_{in}}$ is high: (1) when the button is pressed, the DFFs in this module are reset, driving FN high, which subsequently sets $\overline{RST}$ high, thereby resetting all DFFs in the clock recovery, clock divider, and counter circuits; (2) when the button is released, $\overline{BTN}^{DFF}$ transitions to a high state, causing FN and $\overline{RST}$ to invert, initiating the counter and pulling the CNT signal low; (3) the counter sequentially decrements from 111 to 010, iterating through all six switched capacitors; (4) upon reaching the state 001, CNT, CNT^{DFF}, and FN are inverted, resetting DFFs in all modules except the processing unit. At this stage, the counter becomes inactive, holding CNT high and resulting in a high $\overline{RST}$, which resets the counter. The auto-tuning mechanism reactivates only upon receiving a BTN pulse, thereby enabling the counter. The system utilizes a three-bit down counter and a one-hot decoder to sequentially activate six processing units while regulating the termination process. The counter is constructed using three DFFs, which are reset by $\overline{RST}$, producing a three-bit output $Y\langle 2:0\rangle$. These three bits are decoded into six one-hot selection signals $Sel\langle 5:0\rangle$, ensuring that only one signal is active at any given time. Figure 4.36b provides the truth table correlating $Y\langle 2:0\rangle$ with $Sel\langle 5:0\rangle$. For instance, $Sel\langle 4\rangle$ is determined by the logic expression $Sel\langle 4\rangle = Y\langle 2\rangle \cdot Y\langle 1\rangle \cdot \overline{Y\langle 0\rangle}$. These selection signals interface with the processing unit via AND gates, ensuring that a switched capacitor's control signal is exclusively adjusted within a designated time window. When the counter reaches either the 001 or 000 state, the signal CNT transitions from low to high, governed by the logic expression $CNT = \overline{Y\langle 2\rangle} \cdot \overline{Y\langle 1\rangle}$, signifying the termination of the auto-tuning process.

4.7.4 Clock Recovery and Divider

A square wave is extracted from the AC signal using an inverter, a DFF, and a buffer, as depicted in the clock recovery part of Fig. 4.35. The clock divider, implemented through a cascade of DFFs, generates clock signals at various frequency multiples of the recovered square wave. These clock signals are further processed to produce control signals by selectively combining specific outputs, as illustrated in Fig. 4.35. The reset signal $\overline{RST}$ integrates the universal reset $\overline{RST_{in}}$ with a termination signal, which will be discussed in detail in the following section. Denoting the clock from the nth DFF (among 7 DFFs) of the clock recovery and divider sequence as CLK_{2^n}, where 2^n indicates its period being 2^n times the AC signal period, Φ_1 to Φ_3 are generated as

$$\begin{aligned}\Phi_1 &= \left(CLK_{2^{N-1}} + CLK_{2^{N-2}} + CLK_{2^{N-3}}\right) \cdot CLK_{2^N} \\ \Phi_2 &= \overline{CLK_{2^N} + CLK_{2^{N-1}} + CLK_{2^{N-2}}} \\ \Phi_3 &= \overline{CLK_{2^N} + CLK_{2^{N-1}} + CLK_{2^{N-2}} + CLK_{2^{N-3}}}\end{aligned}$$

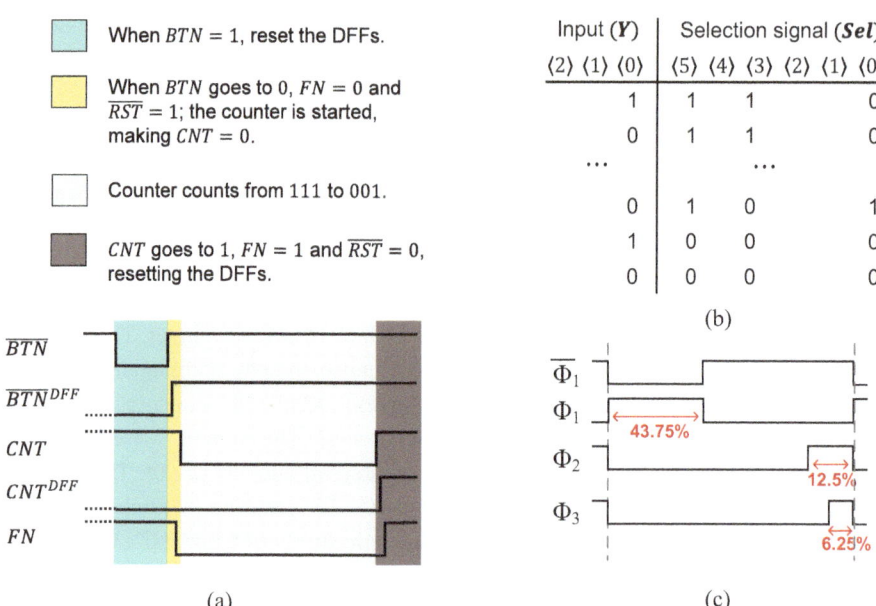

Fig. 4.36 a Startup waveform; b truth table of the one-hot decoder; and c clock signals Φ_1 to Φ_3

where $N = 8$, each with a duty cycle of 43.75, 12.5 and 6.25%, as shown in Fig. 4.36c. Fluctuations may arise when performing an AND operation on clock signals due to the propagation delay of the DFF. For instance, CLK_{2^N} transitions to a low state slightly later than clocks with shorter periods transition to a high state, resulting in a brief interval where both signals remain high simultaneously. However, this transient overlap does not affect the overall functionality of the system.

4.7.5 Control Unit

The control unit operates based on the following strategy: during each of the six periods, the control signal of a switched capacitor is toggled, and the rectified voltage is compared before and after the transition. If the updated rectified voltage is higher, the processing unit keeps this change, as illustrated in the corresponding flowchart. The clocked comparator, responsible for this evaluation, is implemented as shown in Fig. 4.35 [61]. The working mechanism of the control unit is further demonstrated in Fig. 4.37, which provides an example including comparator signals and the corresponding tuning capacitance adjustments. Additionally, the first two periods of the transient simulation are subdivided into Phases (1a) to (2d), with signal variations summarized in Fig. 4.37. This procedure is iterated until all six switched capacitors have been evaluated.

4.7 Automatic Resonance Tuning System

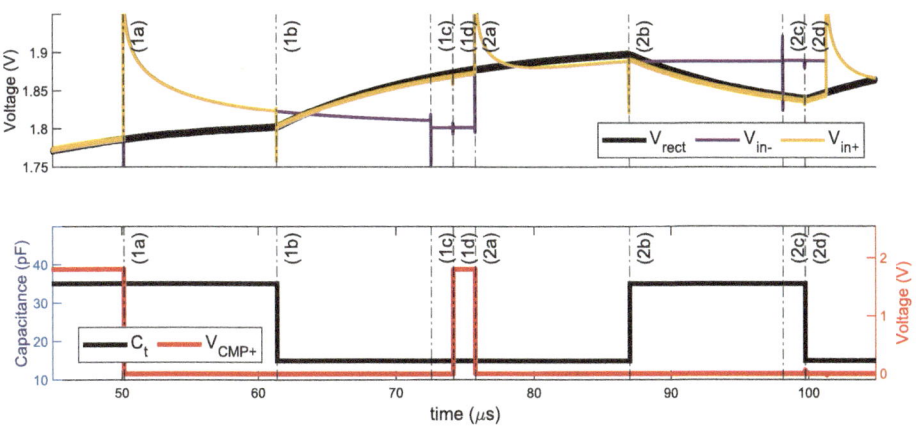

Phase	Variations in the control unit signals
(1a)	Auto-tuning is enabled and $\Phi_1 = 1$; the storage capacitor C_{prev} is connected to rectified voltage V_{rect}; counter counts to 111 and the selection signal $Sel\langle 5\rangle = 1$.
(1b)	$\Phi_1 = 0$ and the storage capacitor is disconnected; the rectified voltage is connected to the positive input of the comparator; the control signal $SW\langle 5\rangle$ is inverted and the connected tuning capacitance changes.
(1c)	$\Phi_2 = 1$ and the storage capacitor is connected to the negative input of the comparator.
(1d)	$\Phi_3 = 1$ and comparison is made; $V_{in+} > V_{in-}$ and V_{CMP+} goes high and the output of the DFF in the 5th processing unit is inverted.
(2a)	Same as (1a) except that the counter counts to 110 and $Sel\langle 4\rangle = 1$.
(2b)	Same as (1b) except that $SW\langle 4\rangle$ is inverted.
(2c)	Same as (1c).
(2d)	Comparison is made and $V_{in+} < V_{in-}$; the output of the DFF in the 4th processing unit is not inverted and the connected tuning capacitance is recovered to the state at (2a).

Fig. 4.37 Simulation of the rectified voltage V_{rect}, input voltages of the comparator V_{in-} and V_{in+}, total tuning capacitance C_t and output voltage of the comparator V_{CMP+} with $C_2 = 390\,\text{pF}$ and $k = 0.1$

At an operating frequency of 13.56 MHz, the off-chip inductors and capacitors on the primary and secondary sides are configured as follows: $L_1 = 1.2\,\mu\text{H}$, $C_1 = 115\,\text{pF}$, $L_2 = 0.6\,\mu\text{H}$, and $C_{2,opt} = 195\,\text{pF}$ (including an on-chip 35 pF capacitor). The switched capacitors $C_{SW}\langle 5:0\rangle$ are specified as 20, 20, 10, 10, 10, and 5 pF, with an initial tuning capacitance of

35 pF, set by inverting the control signals $SW\langle 5\rangle$, $SW\langle 3\rangle$, and $SW\langle 0\rangle$. A binary-weighted capacitor configuration is intentionally avoided to minimize the capacitance transition step size. To ensure signal stability, a buffer consisting of two cascaded inverters is placed at the output of each processing unit, mitigating fluctuations while effectively driving the large transistors controlling the switched capacitors. The clock chain incorporates $N = 8$ D flip-flops, resulting in a delay of 9.6 μs between switching and voltage comparison. The period of Φ_1 is 18.8 μs, leading to a total tuning duration of 153.6 μs.

During the tuning process, the control unit exhibits a simulated power consumption of 9.2 μW, which reduces to 5.8 μW once tuning is completed. The system's performance under a coupling coefficient of $k = 0.1$ is summarized in Fig. 4.38, demonstrating compensation for an LC variation of ±15%. It is important to note that the LC variation is modeled exclusively through capacitance changes. Figure 4.39a presents the transient simulation at 10 MHz with a coupling coefficient of $k = 0.1$, comparing resonance capacitances of 390 and 320 pF. The results indicate a significant enhancement in rectified voltage with the auto-tuning system compared to the untuned case (V_0). Additionally, the adaptive selection of tuning capacitance effectively compensates for deviations in C_2 from its optimal value. The resulting control signals $SW\langle 5:0\rangle$ directly correspond to C_2, such as $SW\langle 5:0\rangle = 000000$ for $C_2 = 390$ pF, $SW\langle 5:0\rangle = 000100$ for $C_2 = 380$ pF, and $SW\langle 5:0\rangle = 111110$ for $C_2 = 330$ pF. A similar performance trend is expected for inductance variations, as the optimal resonance frequency is consistently maintained within the tuning range by adjusting the tuning capacitors. Figure 4.39b provides a detailed breakdown of the simulated power consumption of the logic unit. The clock generation circuitry within the logic unit accounts for the largest share of power consumption. The combined power consumption of the logic unit and the switch array amounts to 113.6 μW. The simulated end-to-end PTE for a coupling coefficient of 0.1 is 25.9%, delivering 6.8 mW to the load.

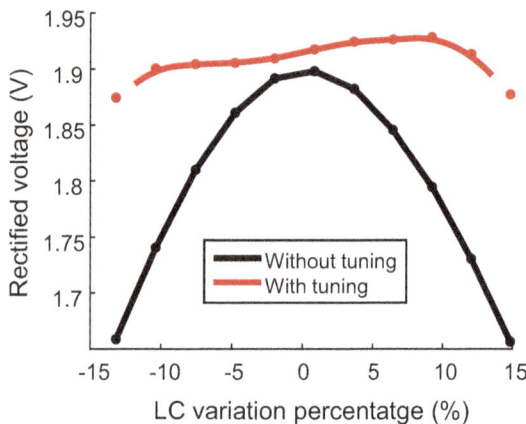

Fig. 4.38 Simulation results of the output voltage of the rectifier without and with the resonance tuning at $k = 0.1$

4.7 Automatic Resonance Tuning System

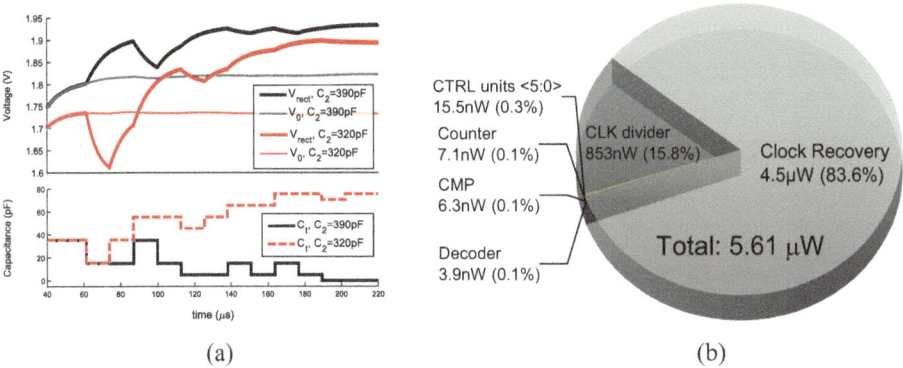

(a)   (b)

Fig. 4.39 Simulation results of **a** the transient rectified voltage with tuning (V_{rect}), rectified voltage without tuning (V_0) and the total added tuning capacitance C_t at $k = 0.1$, **b** power consumption breakdown of the proposed design

4.7.6 Measurement Results

The presented automatic resonance tuning system is developed and fabricated using a 180 nm standard CMOS technology with an area of 0.339 mm² and a core area of 0.1 mm², as shown in Fig. 4.40. A measurement setup is designed to validate the system, including an inductive link, as shown in Fig. 4.41, with the system's connection in Fig. 4.41a. Moreover, a 40 nF off-chip capacitor is used for the output of the rectifier (C_{prev} in Fig. 4.35).

Given the constraint of implanting the Rx coil within a limited outer diameter (approximately 10 mm), while maintaining a constant outer diameter, a trade-off arises between the coil's quality factor and the number of turns per layer (fill ratio). Therefore, optimizing the implant coil design is essential to meet these requirements. The inductive link consists of a Tx coil (1.2 µH) and an Rx implant coil (600 nH), with a separation distance ranging from 5 to 20 mm in air or 1 cm within biological tissue (chicken breast). To enhance PTE, an additional resonator coil can be incorporated into the external unit. For brain implant

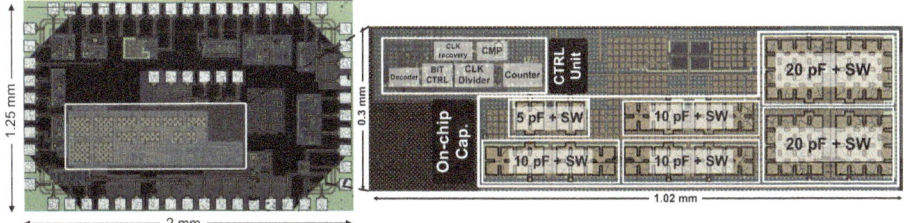

Fig. 4.40 Chip microphotograph of the presented automatic resonance tuning system using a standard 180 nm CMOS process, occupying an area of 1.02 mm × 0.3 mm

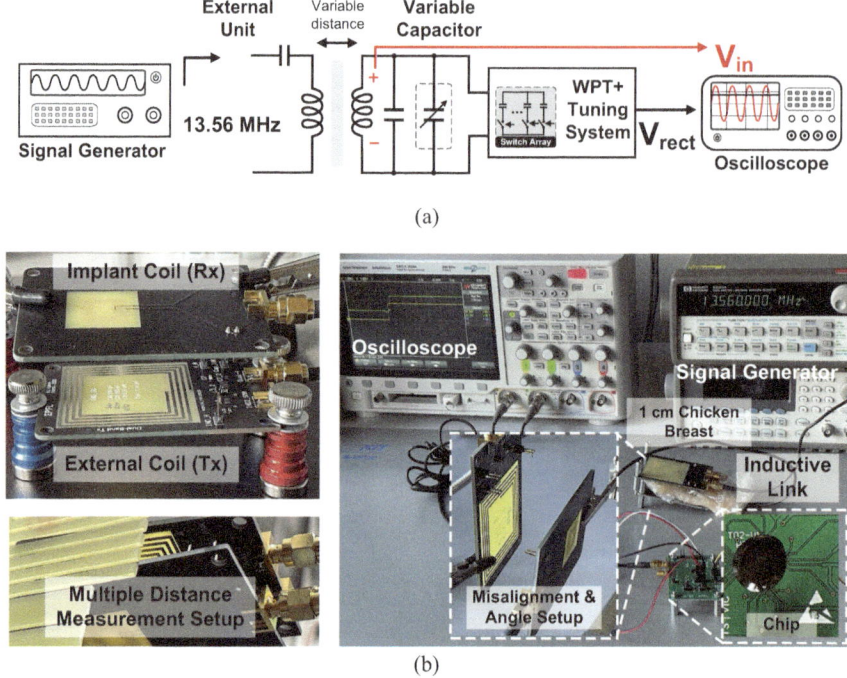

Fig. 4.41 a Schematic and **b** photograph of the experimental setup in air and 1 cm of chicken breast, featuring external and implant coils, along with configurations for multiple distance and angular misalignment tests [60]

applications, the Rx coil's outer diameter is specified as 13.6 mm. Experimental characterization has been performed through transient voltage analysis, as well as distance and angular misalignment sweeps, utilizing the setup illustrated in Fig. 4.41b.

The obtained rectified voltage is measured and evaluated with different LC combinations with positive and negative variations in the resonance frequency to demonstrate the effectiveness of the tuning system. The automatic resonance tuning process is started by using a pre-defined voltage node (BTN), and the rectified voltage (V_{rect}) is measured, as shown in Fig. 4.42. Subsequently, the tuning performance is activated by using the reset button. The measured transient results from both positive and negative LC tank variations are presented in Fig. 4.43. Consequently, the system presents a rectified voltage improvement (and consequently a PTE enhancement).

Figure 4.44a illustrates the impact of the automatic tuning system on the received rectified voltage (normalized) under LC variations, where deviations in the resonance capacitor from its ideal value are observed. The tuning process is reflected in the slope of V_{rect} when tuning is achieved. Without tuning, the rectified voltage exhibits a significant reduction of approximately 7% across the full range of variation. However, this issue is mitigated by the

4.7 Automatic Resonance Tuning System

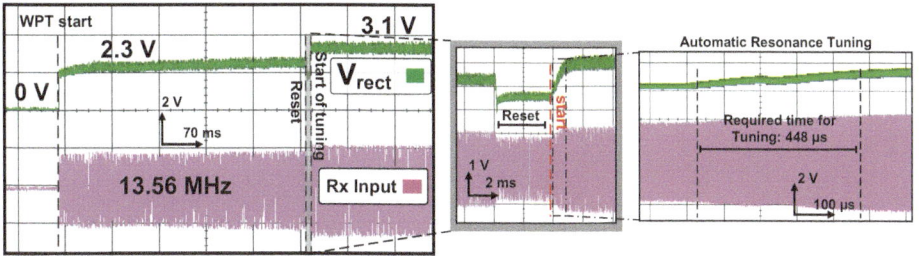

Fig. 4.42 Measured transient received input sinusoidal voltage and the rectified voltage after starting wireless power transfer (WPT) and automatic tuning

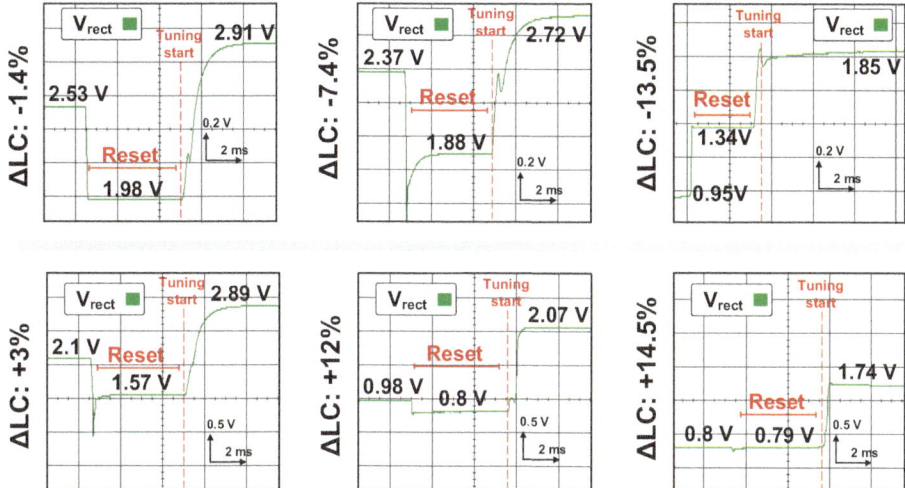

Fig. 4.43 Measured results of transient voltages within automatic tuning at positive and negative LC tank variation

tuning system, which stabilizes the voltage. Additionally, for a 12.2% variation, assuming a constant load, the delivered power to the load is measured at 1.07 mW with tuning and 3.69 mW without tuning, resulting in an increase in PTE from 9.1 to 31.2%. Figure 4.44b presents the relationship between the harvested voltage and the coil separation distance, comparing results with and without the tuning system. A dedicated setup allows for precise adjustments to the angular misalignment at various distances to assess its impact. As depicted in Fig. 4.44c, the automatically tuned system exhibits angular misalignment tolerance of up to 20° at a 10 mm distance. Due to the symmetrical coil configuration, sensitivity to Z-axis misalignment is negligible, while the sensitivity to X- and Y-axis misalignment remains similar.

The transient response of the rectified voltage under varying coil separation distances, with and without tuning, is shown in Fig. 4.45. In this experimental setup, the Tx coil is fixed

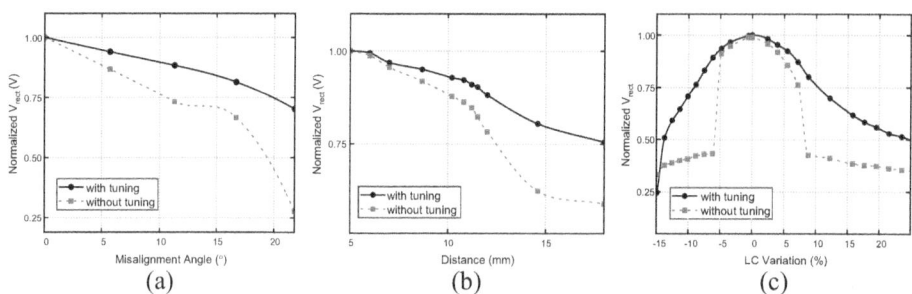

Fig. 4.44 Measurement results of **a** the normalized V_{rect} without and with resonance tuning in different LC tank variations, at no LC variation, and the received rectifier voltage versus **b** distance and **c** angular misalignment

Fig. 4.45 Measured transient rectified voltage results with and without automatic resonance tuning under distance variation from 0 to 10 cm

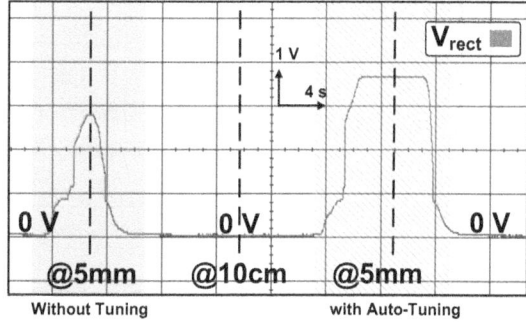

in place while the Rx coil is gradually displaced from an initial distance of 10 cm to a final position of 5 mm. The tuning system is then activated, and the entire process is repeated under the adjusted conditions. A precision twisting screw jack is utilized to ensure uniform displacement and maintain a consistent motion speed throughout the experiment.

4.7.7 Discussion

The tuning control unit, including the switch array and clock generation modules, consumes 154.7 µW. The overall system delivers between 1–10 mW of power, achieving a maximum end-to-end PTE of 31.2% and an inductive link efficiency of 48.9%. A summary of the automatic resonance tuning system and its wireless power conversion performance is provided in Table 4.4. Additionally, Table 4.5 presents a comparative analysis of the proposed auto-tuning system against state-of-the-art tuning approaches and compensation techniques, highlighting key performance metrics, tuning methodologies, and resolution. Automatic tuning systems are classified based on the tuning strategy-either SAR or monotonic sweeping-as well as the detection mechanism, which may rely on rectified voltage/current measurements

Table 4.4 Auto-tuning performance summary

CMOS technology	TSMC 180 nm
Chip area	0.339 mm^2
Operation frequency	13.56 MHz or 6.78 MHz
Wireless power transmission 1 external + 1 implant	
Coil distance	5–20 mm
Inductive link: L_{Tx}	L_{Rx}	1.2 μH	0.6 μH
Resonance capacitor C_{Tx}	C_{Rx}	115 pF	230 pF
Automatic resonance tuning system	
Total on-chip cap.	75 pF
Resolution	5 pF
Number of tuning bits	6 control bits
Tuning method	Monotonic sweep
Tuning range	±15%
Power consumption	154.7 μW
Operation frequency	53 kHz

or swing amplitude detection. The proposed wireless power conversion and auto-tuning system offer an efficient on-chip solution in terms of both power consumption and response time, effectively compensating for wide LC tank variations in biomedical implants. The integration of on-chip tuning capacitors facilitates seamless system-on-chip implementation. However, this approach requires a larger silicon footprint compared to tuning systems that utilize off-chip tuning capacitors.

4.8 Summary

In this chapter, the design and experimental results of a wireless power conversion chain and transmission platform were presented. Various circuit and system design techniques were demonstrated that enable efficient remote powering using a single inductive link in the implant. Also, the key components for wireless power conversion are presented, including:

- A digitally-assisted active rectifier with a novel structure that achieves high PCE (80%) and VCR (93%) during rectification.
- A 1.2 V CMOS voltage reference with two versions of LDOs (1.2 and 1.8 V). These circuits are designed from established architectural models in the literature, modified with improvements and customizations tailored to the specific requirements of the project.

Table 4.5 Comparative analysis of automatic resonance tuning systems from existing literature [60]

	ISCAS' 13 [49]	TBCAS' 16 [52]	ISCAS' 17 [48]	JSSC' 20 [53]	TBCAS' 21 [47]	This work
CMOS tech. (μm)	0.35	0.35	0.180	0.065	0.18	0.18
Power frequency (MHz)	13.56	13.56	13.56	13.56	13.56	13.56
Silicon measurement	No	Yes	No	Yes	Yes	Yes
Tuning method	Monotonic sweeping	Monotonic sweeping	SAR	Monotonic sweeping	SAR	Monotonic sweeping
Tuning resolution (pF)	0.51	0.5	0.48	2	0.38	5
# of tuning bits	3-bit	On-chip: 5-bit 0–16 pF Off-chip: 5-bit 15–480 pF	4-bit	6 CNTRL bits 0–128 pF	6 CNTRL bits 0–24 pF	6 CNTRL bits 0–75 pF
Detection target	Rectified current	Swing amplitude	Rectified voltage	Rectified current	Swing amplitude	Rectified voltage
Operation frequency		26.4 kHz	26.5 kHz	100 kHz	5 kHz	39 kHz
Rx coil inductance Outer diameter	2.92 μH	0.6 μH 34 mm	4.59 μH	7.45 μH 22 mm	6.36 μH 12.1 mm	0.6 μH 13.6 mm
LC variation tolerance Max. ΔC coverage	–	6.9% 16 pF	$\pm$11% (overall 22%) 7.2 pF	128 pF	24 pF	$\pm$15% (overall 30%) 75 pF
Tx-Rx distance (mm)	–	20 mm	–	–	20 mm	5–20 mm
Rx power Consumption (mW)	–	Tuning: 224 μW Total: 7.84[a] mW	–	–	Total Rx: 64.6[a] μW	Tuning: 154.7 μW Total Rx: 1.95 mW
Rx efficiency (%)	–	80.3[a]	–	75.4	46.4	64
Delivered power (mW)	–	32	0.85	10	0.03	1–10
Silicon area (mm^2)	–	2.54	0.126	0.22[a]	0.17[a]	0.339
Rectifier type and PCE	–	Active PCE: 76.2%	Passive	Active PCE: 75.4%	Passive	Active PCE: 80.2%

[a] Calculated value

- A power control unit including power feedback, voltage limiter, and a power-on-reset unit.
- A wireless power transfer unit including a relaxation oscillator with a well-established architecture of a class-E power amplifier and a buck DC-DC converter.
- A novel on-chip automatic resonance tuning system designed to efficiently tune resonance at the desired frequency for the implant.

The ASIC was designed and implemented using a standard 180 nm CMOS process. Separate measurements were conducted to characterize the performance of the power conversion chain and transfer platform.

References

1. M. J. Karimi, A. Schmid, and C. Dehollain, "Wireless Power and Data Transmission for Implanted Devices via Inductive Links: A Systematic Review," *IEEE Sensors Journal*, vol. 21, no. 6, pp. 7145–7161, 2021.
2. Y. Jia, S. A. Mirbozorgi, P. Zhang, O. T. Inan, W. Li, and M. Ghovanloo, "A Dual-Band Wireless Power Transmission System for Evaluating mm-Sized Implants," *IEEE Transactions on Biomedical Circuits and Systems*, vol. 13, pp. 595–607, 2019.
3. P. D. Wolf, "Thermal considerations for the design of an implanted cortical brain-machine interface (BMI)," *Indwelling Neural Implants: Strategies for Contending with the in Vivo Environment*, vol. 3, pp. 63–86, 2007.
4. P. Yeon, S. A. Mirbozorgi, J. Lim, and M. Ghovanloo, "Feasibility Study on Active Back Telemetry and Power Transmission Through an Inductive Link for Millimeter-Sized Biomedical Implants," *IEEE Transactions on Biomedical Circuits and Systems*, vol. 11, pp. 1366–1376, 2017.
5. M. J. Karimi, M. Jin, Y. Zhou, C. Dehollain, and A. Schmid, "Wirelessly Powered and Bi-directional Data Communication System with Adaptive Conversion Chain for Multisite Biomedical Implants Over Single Inductive Link," *IEEE Transactions on Biomedical Circuits and Systems*, pp. 1–11, 2024.
6. Y.-H. Lam, W.-H. Ki, and C.-Y. Tsui, "Integrated Low-Loss CMOS Active Rectifier for Wirelessly Powered Devices," *IEEE Transactions on Circuits and Systems II: Express Briefs*, vol. 53, no. 12, pp. 1378–1382, 2006.
7. M. J. Karimi, C. Dehollain, and A. Schmid, "An offset-enhanced active rectifier with delay compensated active diodes for wirelessly powered biomedical implants," *Electronics Letters*, vol. 60, no. 15, p. e13274, 2024.
8. Y. Lu and W.-H. Ki, "A 13.56 MHz CMOS Active Rectifier With Switched-Offset and Compensated Biasing for Biomedical Wireless Power Transfer Systems," *IEEE Transactions on Biomedical Circuits and Systems*, vol. 8, no. 3, pp. 334–344, 2014.
9. Z. Xue, S. Fan, D. Li, L. Zhang, W. Gou, and L. Geng, "A 13.56 MHz, 94.1% Peak Efficiency CMOS Active Rectifier With Adaptive Delay Time Control for Wireless Power Transmission Systems," *IEEE Journal of Solid-State Circuits*, vol. 54, no. 6, pp. 1744–1754, 2019.
10. S. Shahsavari and M. Saberi, "A Power-Efficient CMOS Active Rectifier with Circuit Delay Compensation for Wireless Power Transfer Systems," *Circuits, Systems, and Signal Processing*, vol. 38, no. 3, pp. 947–966, 2019.

11. G. Namgoong, E. Choi, W. Park, B. Lee, H. Park, H. Ma, and F. Bien, "A 6.78 MHz, 95.0% Peak Efficiency Monolithic Two-Dimensional Calibrated Active Rectifier for Wirelessly Powered Implantable Biomedical Devices," *IEEE Transactions on Biomedical Circuits and Systems*, vol. 15, no. 3, pp. 509–521, 2021.
12. M. J. Karimi, C. Dehollain, and A. Schmid, "A 13.56 MHz Active Rectifier with Digitally-Assisted and Delay Compensated Comparators for Biomedical Implantable Devices," in *ESSCIRC 2023- IEEE 49th European Solid State Circuits Conference (ESSCIRC)*, pp. 313–316, 2023.
13. L. Cheng, W.-H. Ki, Y. Lu, and T.-S. Yim, "Adaptive On/Off Delay-Compensated Active Rectifiers for Wireless Power Transfer Systems," *IEEE Journal of Solid-State Circuits*, vol. 51, no. 3, pp. 712–723, 2016.
14. Y. Lu and W.-H. Ki, *CMOS Integrated Circuit Design for Wireless Power Transfer*. Singapore: Springer, 2018.
15. K. Noh, J. Amanor-Boadu, M. Zhang, and E. Sánchez-Sinencio, "A 13.56-MHz CMOS Active Rectifier With a Voltage Mode Switched-Offset Comparator for Implantable Medical Devices," *IEEE Transactions on Very Large Scale Integration (VLSI) Systems*, vol. 26, no. 10, pp. 2050–2060, 2018.
16. R. Erfani, F. Marefat, S. Nag, and P. Mohseni, "A 1–10-MHz Frequency-Aware CMOS Active Rectifier With Dual-Loop Adaptive Delay Compensation and >230-mW Output Power for Capacitively Powered Biomedical Implants," *IEEE Journal of Solid-State Circuits*, vol. 55, no. 3, pp. 756–766, 2020.
17. L. Cheng, X. Ge, L. Hu, Y. Yao, W.-H. Ki, and C.-Y. Tsui, "A 40.68-MHz Active Rectifier With Hybrid Adaptive On/Off Delay-Compensation Scheme for Biomedical Implantable Devices," *IEEE Transactions on Circuits and Systems I: Regular Papers*, vol. 67, no. 2, pp. 516–525, 2020.
18. Z. Luo, J. Liu, and H. Lee, "A 40.68-MHz Active Rectifier With Cycle-Based On-/Off-Delay Compensation for High-Current Biomedical Implants," *IEEE Journal of Solid-State Circuits*, vol. 58, no. 2, pp. 345–356, 2023.
19. C. Miroslav, "Design of low-dropout voltage regulator," Master's thesis, Czech Technical University in Prague, 2016.
20. J. Pérez-Bailón, B. Calvo, and N. Medrano, "A Fully-Integrated 180 nm CMOS 1.2 V Low-Dropout Regulator for Low-Power Portable Applications," *Electronics*, vol. 10, no. 17, 2021.
21. S. Roy, A. N. M. W. Azad, S. Baidya, M. K. Alam, and F. Khan, "Powering Solutions for Biomedical Sensors and Implants Inside the Human Body: A Comprehensive Review on Energy Harvesting Units, Energy Storage, and Wireless Power Transfer Techniques," *IEEE Transactions on Power Electronics*, vol. 37, no. 10, pp. 12237–12263, 2022.
22. I. Habibagahi, R. P. Mathews, A. Ray, and A. Babakhani, "Design and Implementation of Multisite Stimulation System Using a Double-Tuned Transmitter Coil and Miniaturized Implants," *IEEE Microwave and Wireless Technology Letters*, vol. 33, no. 3, pp. 351–354, 2023.
23. K. M. Silay, C. Dehollain, and M. Declercq, "A Closed-Loop Remote Powering Link for Wireless Cortical Implants," *IEEE Sensors Journal*, vol. 13, no. 9, pp. 3226–3235, 2013.
24. J. Pan, A. A. Abidi, W. Jiang, and D. Marković, "Simultaneous Transmission of Up To 94-mW Self-Regulated Wireless Power and Up To 5-Mb/s Reverse Data Over a Single Pair of Coils," *IEEE Journal of Solid-State Circuits*, vol. 54, no. 4, pp. 1003–1016, 2019.
25. C. Kim, S. Ha, J. Park, A. Akinin, P. P. Mercier, and G. Cauwenberghs, "A 144-MHz Fully Integrated Resonant Regulating Rectifier With Hybrid Pulse Modulation for mm-Sized Implants," *IEEE Journal of Solid-State Circuits*, vol. 52, no. 11, pp. 3043–3055, 2017.
26. Y. Chen, Y. Luo, and J. Guo, "A 1-W, 6.78-MHz Wireless Power Transfer System With Up-to-16.1% Light-Load Efficiency Improvement and Instant Response Through Single-Cycle-Based DTX Control," *IEEE Transactions on Power Electronics*, vol. 38, no. 2, pp. 2743–2753, 2023.

References

27. S. Hao and S. Taylor, "A closed-loop inductive power control system for an instrumented strain sensing tibial implant," in *2014 36th Annual International Conference of the IEEE Engineering in Medicine and Biology Society*, pp. 6553–6556, 2014.
28. A. Rashidi, K. Laursen, S. Hosseini, and F. Moradi, "Overvoltage Protection Circuits for Ultrasonically Powered Implantable Microsystems," in *2019 41st Annual International Conference of the IEEE Engineering in Medicine and Biology Society (EMBC)*, pp. 4354–4358, 2019.
29. M. Kiani, B. Lee, P. Yeon, and M. Ghovanloo, "A power-management ASIC with Q-modulation capability for efficient inductive power transmission," in *2015 IEEE International Solid-State Circuits Conference - (ISSCC) Digest of Technical Papers*, pp. 1–3, 2015.
30. S.-Y. Lee, C.-H. Hsieh, and C.-M. Yang, "Wireless Front-End With Power Management for an Implantable Cardiac Microstimulator," *IEEE Transactions on Biomedical Circuits and Systems*, vol. 6, no. 1, pp. 28–38, 2012.
31. E. G. Kilinc, C. Baj-Rossi, S. Ghoreishizadeh, S. Riario, F. Stradolini, C. Boero, G. De Micheli, F. Maloberti, S. Carrara, and C. Dehollain, "A System for Wireless Power Transfer and Data Communication of Long-Term Bio-Monitoring," *IEEE Sensors Journal*, vol. 15, pp. 6559–6569, 2015.
32. G. Balachandran and R. Barnett, "A 110 nA Voltage Regulator System With Dynamic Bandwidth Boosting for RFID Systems," *IEEE Journal of Solid-State Circuits*, vol. 41, no. 9, pp. 2019–2028, 2006.
33. M. J. Karimi, C. Dehollain, and A. Schmid, "Power Feedback Control Unit for Closed-Loop Wirelessly Powered Biomedical Implants," *IEEE Transactions on Circuits and Systems II: Express Briefs*, vol. 70, no. 5, pp. 1674–1678, 2023.
34. L. Cheng, W.-H. Ki, and C.-Y. Tsui, "A 6.78-MHz Single-Stage Wireless Power Receiver Using a 3-Mode Reconfigurable Resonant Regulating Rectifier," *IEEE Journal of Solid-State Circuits*, vol. 52, no. 5, pp. 1412–1423, 2017.
35. C. Kim, J. Park, S. Ha, A. Akinin, R. Kubendran, P. P. Mercier, and G. Cauwenberghs, "A 3 mm*3 mm Fully Integrated Wireless Power Receiver and Neural Interface System-on-Chip," *IEEE Transactions on Biomedical Circuits and Systems*, vol. 13, pp. 1736–1746, 2019.
36. P. Feng and T. G. Constandinou, "Autonomous Wireless System for Robust and Efficient Inductive Power Transmission to Multi-Node Implants," in *2021 IEEE International Symposium on Circuits and Systems (ISCAS)*, pp. 1–5, 2021.
37. F.-B. Yang, J. Fuh, Y.-H. Li, M. Takamiya, and P.-H. Chen, "Structure-Reconfigurable Power Amplifier (SR-PA) and 0X/1X Regulating Rectifier for Adaptive Power Control in Wireless Power Transfer System," *IEEE Journal of Solid-State Circuits*, vol. 56, no. 7, pp. 2054–2064, 2021.
38. G. Zhao, Z. Xiao, P.-I. Mak, R. P. Martins, and M.-K. Law, "One-Cycle-Startup Relaxation Oscillator Using Ratiometric Threshold-Referenced and Self-Synchronized Power Gating Techniques," *IEEE Transactions on Circuits and Systems II: Express Briefs*, pp. 1–1, 2023.
39. H. Wang and P. P. Mercier, "A 763 pW 230 pJ/Conversion Fully Integrated CMOS Temperature-to-Digital Converter With 0.81°C /-0.75° C Inaccuracy," *IEEE Journal of Solid-State Circuits*, vol. 54, no. 8, pp. 2281–2290, 2019.
40. M. J. Karimi, S. Mehdi, C. Dehollain, and A. Schmid, "Wireless Power and Data Transceiver in A Central Implanted Unit for Biomedical Applications," in *2024 IEEE 15th Latin America Symposium on Circuits and Systems (LASCAS)*, pp. 1–5, 2024.
41. N. Sokal and A. Sokal, "Class E-A new class of high-efficiency tuned single-ended switching power amplifiers," *IEEE Journal of Solid-State Circuits*, vol. 10, p. 168–176, 1975.
42. M. M. Ahmadi, S. Pezeshkpour, and Z. Kabirkhoo, "A High-Efficiency ASK-Modulated Class-E Power and Data Transmitter for Medical Implants," *IEEE Transactions on Power Electronics*, vol. 37, no. 1, pp. 1090–1101, 2022.

43. M. M. Ahmadi and M. Sarbandi-Farahani, "A Class-E Power and Data Transmitter with On-Off Keying Data Modulation for Wireless Power and Data Transmission to Medical Implants," *Circuits, Systems, and Signal Processing*, vol. 39, p. 4174–4186, 2020.
44. C. F. Lee and P. Mok, "A monolithic current-mode CMOS DC-DC converter with on-chip current-sensing technique," *IEEE Journal of Solid-State Circuits*, vol. 39, no. 1, pp. 3–14, 2004.
45. M. W. Baker and R. Sarpeshkar, "Feedback analysis and design of RF power links for low-power bionic systems," *IEEE Transactions on Biomedical Circuits and Systems*, vol. 1, pp. 28–38, 2007.
46. R.-F. Xue, K.-W. Cheng, and M. Je, "High-Efficiency Wireless Power Transfer for Biomedical Implants by Optimal Resonant Load Transformation," *IEEE Transactions on Circuits and Systems I: Regular Papers*, vol. 60, no. 4, pp. 867–874, 2013.
47. H. Lyu and A. Babakhani, "A 13.56-MHz -25-dBm-Sensitivity Inductive Power Receiver System-on-a-Chip With a Self-Adaptive Successive Approximation Resonance Compensation Front-End for Ultra-Low-Power Medical Implants," *IEEE Transactions on Biomedical Circuits and Systems*, vol. 15, no. 1, pp. 80–90, 2021.
48. P. Gosselin, R. Puddu, A. Carreira, M. Ghanad, M. Barbaro, and C. Dehollain, "A CMOS automatic tuning system to maximize remote powering efficiency," in *2017 IEEE International Symposium on Circuits and Systems (ISCAS)*, pp. 1–4, 2017.
49. H. Xu, U. Bihr, J. Becker, and M. Ortmanns, "A multi-channel neural stimulator with resonance compensated inductive receiver and closed-loop smart power management," in *2013 IEEE International Symposium on Circuits and Systems (ISCAS)*, pp. 638–641, 2013.
50. H.-M. Lee and M. Ghovanloo, "A Power-Efficient Wireless Capacitor Charging System Through an Inductive Link," *IEEE Transactions on Circuits and Systems II: Express Briefs*, vol. 60, no. 10, pp. 707–711, 2013.
51. H.-M. Lee, K. Y. Kwon, W. Li, and M. Ghovanloo, "A Power-Efficient Switched-Capacitor Stimulating System for Electrical/Optical Deep Brain Stimulation," *IEEE Journal of Solid-State Circuits*, vol. 50, no. 1, pp. 360–374, 2015.
52. B. Lee, D. Ahn, and M. Ghovanloo, "Three-Phase Time-Multiplexed Planar Power Transmission to Distributed Implants," *IEEE Journal of Emerging and Selected Topics in Power Electronics*, vol. 4, pp. 263–272, 2016.
53. D. Ye, Y. Wang, Y. Xiang, L. Lyu, H. Min, and C. J. Shi, "A Wireless Power and Data Transfer Receiver Achieving 75.4% Effective Power Conversion Efficiency and Supporting 0.1% Modulation Depth for ASK Demodulation," *IEEE Journal of Solid-State Circuits*, vol. 55, no. 5, pp. 1386–1400, 2020.
54. S. Lee, J. Yoo, H. Kim, and H.-J. Yoo, "A dynamic real-time capacitor compensated inductive coupling transceiver for wearable body sensor network," in *2009 Symposium on VLSI Circuits*, pp. 42–43, 2009.
55. S. O'Driscoll, A. S. Y. Poon, and T. H. Meng, "A mm-sized implantable power receiver with adaptive link compensation," in *2009 IEEE International Solid-State Circuits Conference - Digest of Technical Papers*, pp. 294–295, 295a, 2009.
56. M. Stoopman, S. Keyrouz, H. J. Visser, K. Philips, and W. A. Serdijn, "Co-Design of a CMOS Rectifier and Small Loop Antenna for Highly Sensitive RF Energy Harvesters," *IEEE Journal of Solid-State Circuits*, vol. 49, no. 3, pp. 622–634, 2014.
57. Y. An, X. Li, X. Feng, H. Xu, and Y. Zhuang, "High Sensitivity RF Energy Harvesting System with Self-calibrate Network," in *2023 IEEE/MTT-S International Microwave Symposium - IMS 2023*, pp. 875–878, 2023.
58. S.-H. Lee, Y.-W. Jeong, S.-J. Park, and S.-U. Shin, "A Rectifier-Reusing Bias-Flip Energy Harvesting Interface Circuit With Adaptively Reconfigurable SC Converter for Wind-Driven Triboelectric Nanogenerator," *IEEE Transactions on Industrial Electronics*, vol. 70, no. 8, pp. 8022–8031, 2023.

59. J. Pan, A. A. Abidi, W. Jiang, and D. Marković, "Simultaneous Transmission of Up To 94-mW Self-Regulated Wireless Power and Up To 5-Mb/s Reverse Data Over a Single Pair of Coils," *IEEE Journal of Solid-State Circuits*, vol. 54, no. 4, pp. 1003–1016, 2019.
60. M. J. Karimi, M. Jin, C. Dehollain, and A. Schmid, "A Wireless Power Conversion Chain With Fully On-Chip Automatic Resonance Tuning System for Biomedical Implants," *IEEE Open Journal of Circuits and Systems*, vol. 5, pp. 117–127, 2024.
61. M. Abbas, Y. Furukawa, S. Komatsu, J. Y. Takahiro, and K. Asada, "Clocked comparator for high-speed applications in 65nm technology," in *2010 IEEE Asian Solid-State Circuits Conference*, pp. 1–4, 2010.

Wireless Data Communication 5

In the context of brain-computer interface (BCI) systems, capturing a substantial number of channel signals and sending them to an external unit is necessary. Receiving commands and configuration bits is also essential for a base station or an external unit [1]. Various wireless data communication methods are proposed that can be categorized into two groups. One uses wireless power transfer parameters for data communication on the same magnetic wave which is suitable for remotely powered systems. The alternative option is to design separate transceivers and coils on the implant and external unit only for data communication. The choice depends on factors such as the power budget, data rate, and available antenna space. In this chapter, a low-power data communication system is presented. The integration of both wireless power and data transmission into medical implants presents several challenges, including limited size and low power efficiency of the power link, as well as robust data link requirements. In the downlink data path, the ASK modulation scheme is commonly used for low-power medical implants [2–5]. For uplink communication, LSK is a convenient method in which data is transferred by adjusting the load on the LC tank at the receiver. It disrupts the resonant condition of the inductive link and changes the voltage at the Tx terminal. LSK offers the advantage that the ASK demodulator can be reused to demodulate the LSK signal, as the received LSK signal is similar to the ASK signal. The design of data communication units is studied in the following, including the clock recovery, modulator, and ASK and FSK demodulators. The schematic of the DCC unit and its timing diagram are shown in Fig. 5.1.

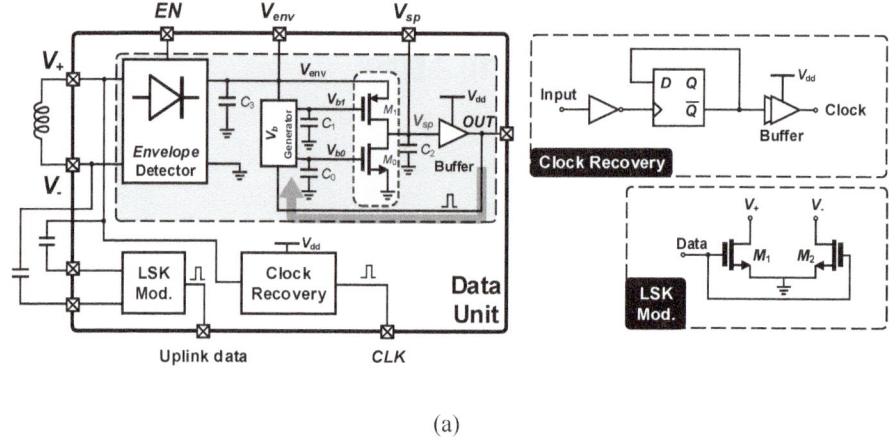

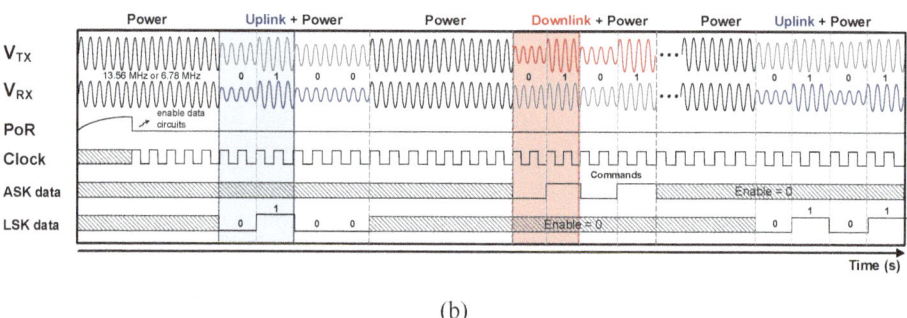

Fig. 5.1 **a** Schematic of the data conversion chain (DCC) unit and **b** its timing diagram

5.1 Clock Recovery (CR)

In implant electronics, an internal clock plays a crucial role in precisely sampling the recorded signals that are obtained. Hence, an oscillator in the chip or a clock recovery circuit is needed. The latter is designed for the sake of simplicity for the ASPs. Figure 5.1a shows the schematic of the proposed clock recovery circuit. It comprises an inverter and a master-slave D flip-flop (DFF) configuration. The inverter generates a square wave at half the frequency of operation, and the use of a DFF ensures a 50% duty cycle of the generated square wave.

5.2 LSK Modulator

In the proposed system, uplink communication is facilitated using LSK, where data is transmitted by intentionally disrupting the resonant condition of the inductive link through load variation in the LC tank at the receiver end. This disruption induces a voltage change at the transmitter (Tx) terminal. It is important to note that the auto-tuning mechanism must be deactivated during LSK modulation to ensure proper operation. The LSK technique enables data transmission over the same magnetic field used for power delivery, offering an efficient approach to reverse telemetry. Data modulation is accomplished by controlling switches, as depicted in Fig. 5.1a, which alter the resonant characteristics of the LC tank, thereby modulating the voltage at the terminals of the primary coil.

5.3 Separated-Vb ASK Demodulator

The downlink data demodulator must be compact and efficient, as it is integrated within the implant. Given its simplicity, ASK is one of the most widely adopted modulation schemes for low-power IMDs. Key design parameters include the modulation index (MI) [2, 6], carrier frequency, power consumption, and chip area. While a lower modulation index improves power link efficiency, it also reduces the ability to distinguish symbols, compromising robustness. Several ASK demodulator topologies have been proposed to optimize these parameters. The work in [7] introduces a design utilizing two resistors and capacitors with different time constants for envelope signal extraction. Further minimization of passive components has been explored to reduce chip area, as demonstrated in [8], where only one resistor is used. In [9], the resistor is replaced with a transistor operating in the linear region. The approach in [10] employs a voltage shifter after the rectifier to generate a low-voltage envelope replica signal, while an averaging circuit produces a threshold voltage reference for output detection. Another topology, proposed in [11, 12], utilizes diode-connected transistors and a MOSCAP for envelope detection, followed by a digital shaper that provides a self-bias voltage. However, this circuit architecture is ineffective at low modulation indices due to its inherent design constraints. To overcome these limitations, a novel ASK demodulator based on separately biased digital shaping of the envelope signal is introduced, as illustrated in Fig. 5.2. This design enhances circuit flexibility and reduces demodulation delay. Furthermore, the circuit operates with low power consumption due to its switching-based architecture and the absence of a comparator.

5.3.1 Circuit Architecture

This part presents a detailed analysis of the self-sampling ASK demodulator. A key component of this demodulator is the digital shaper, whose core structure is depicted in the left part

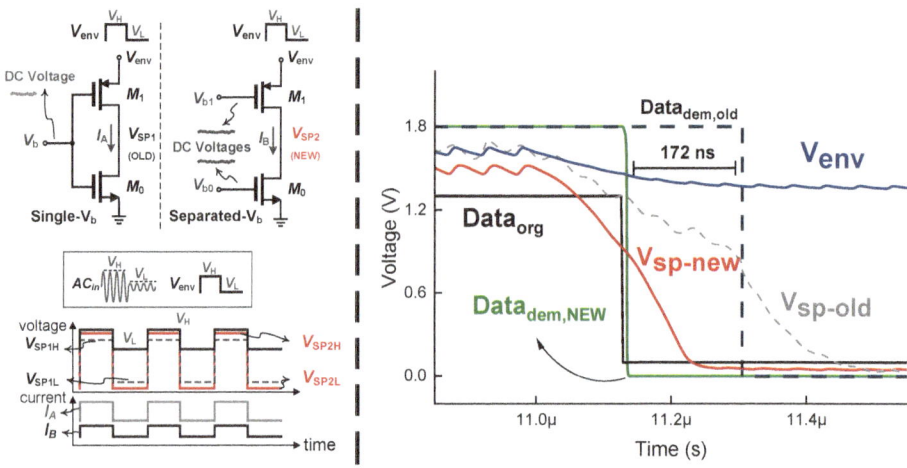

Fig. 5.2 Simplified schematic of the self-sampling ASK demodulator core

of Fig. 5.2 [13]. The envelope signal, V_{env}, alternates between two voltage levels, V_H and V_L, representing binary data '1' and '0', respectively. The primary objective of this circuit is to establish $V_{SP} = V_{SP_H}$ when $V_{env} = V_H$ and $V_{SP} = V_{SP_L}$ when $V_{env} = V_L$. To optimize performance, $\Delta V_{env} = V_H - V_L$ must be minimized, whereas $\Delta V_{SP} = V_{SP_H} - V_{SP_L}$ must be maximized to enhance level distinction and reduce the required modulation index. To achieve this, the circuit is designed such that when $V_{env} = V_H$, the output voltage is maximized. It can be analytically demonstrated that the optimal biasing condition requires transistor M_0 to operate in the saturation region while M_1 functions in the linear region. Conversely, when $V_{env} = V_L$, M_0 transitions to the linear region while M_1 operates in saturation, ensuring that V_{SP_L} remains at its minimum. These operational insights are derived through comprehensive circuit equations considering various transistor operating regions. Assuming that V_b is constant, the objective is to determine an appropriate bias voltage V_b along with the optimal transistor width-to-length ratios, $(W/L)_{0,1}$. When $V_{env} = V_H$, M_0 operates in saturation, and M_1 is in the linear region, with the following relationships: $V_{GS_0} = V_b$, $V_{SG_1} = V_H - V_b$, $V_{SD_1} = V_H - V_{SPH}$, where $V_{th_{n,p}}$ denotes the threshold voltages of the NMOS and PMOS transistors. In the subsequent state, when $V_{env} = V_L$, transistor M_0 transitions to the linear region while M_1 enters saturation, leading to the conditions: $V_{GS_0} = V_b$, $V_{DS_0} = V_{SP_L}$, and $V_{SG_1} = V_L - V_b$. Considering $I_{M0} = I_{M1}$ in the two states, $V_{SPH,L}$ can be calculated based on V_b and the transistor dimensions:

$$\Delta V_{SP} = V_{SPH} - V_{SPL}, \quad i_s = \frac{I_{S1}}{I_{S0}} = \frac{k'_p \left(\frac{W}{L}\right)_1}{k'_n \left(\frac{W}{L}\right)_0} \tag{5.1}$$

where i_s is the ratio of the constant of technology and the dimension of M_1 to M_0, and $k'_{n,p}$ are NMOS/PMOS process transconductance parameters. To obtain boundary conditions

5.3 Separated-Vb ASK Demodulator

between saturation and linear regions, $V_{DS} = V_{GS} - V_{th}$ is assumed for both states, yielding Eq. (5.2).

$$\frac{k'_n}{2}\left(\frac{W}{L}\right)_0 (V_b - V_{thn})^2 < \frac{k'_p}{2}\left(\frac{W}{L}\right)_1 (V_H - V_b - |V_{thp}|)^2$$
$$\frac{k'_p}{2}\left(\frac{W}{L}\right)_1 (V_L - V_b - |V_{thp}|)^2 < \frac{k'_n}{2}\left(\frac{W}{L}\right)_0 (V_b - V_{thn})^2 \quad (5.2)$$

Finally, summing up the terms in (5.2) yields Eq. (5.3):

$$i_s(V_L - V_b - |V_{thp}|)^2 < (V_b - V_{thn})^2 < i_s(V_H - V_b - |V_{thp}|)^2 \quad (5.3)$$

The designed circuit must function reliably across different envelope voltage levels and modulation indices. Therefore, robustness simulations are essential to evaluate system stability under varying conditions. The simulation results reveal that the constraint on V_b for V_H values ranging from 1 to 4 V depends significantly on the modulation indices (0.03, 0.05, 0.1, and 0.5). The bias voltage (V_b), normalized to V_H, exhibits feasible solutions within a region defined by shifting boundaries influenced by factors such as the modulation index and V_H. This defines a suitable operating region for the ASK demodulator across all scenarios. Notably, as the modulation index increases, the feasible region expands. For these simulations, the high-level voltage (V_H) is assumed to be 2.2 V, while the low-level voltage (V_L) is set to 1.8 V. The current ratio parameter, i_s, varies between 1 and 100, while the range of V_b is constrained by Eq. (5.3). When V_b satisfies this constraint, M_0 (or M_1) operates in the linear region when $V_{env} = V_L$ (or $V_{env} = V_H$), ensuring a broad range of ΔV_{SP}. Additionally, a clear relationship emerges between V_b and i_s when ΔV_{SP} reaches a local maximum. The valid range of V_b expands as i_s increases and V_L decreases. Higher values of i_s and lower V_L also contribute to an increase in the maximum ΔV_{SP}. A broader valid range of V_b translates to improved robustness of the device. Consequently, to enhance system reliability, a higher i_s and a lower V_L are required.

When the modulation index is low (below 0.1), the feasible operating region becomes significantly constrained, allowing functionality only under specific voltage conditions. To enhance circuit performance and expand design flexibility, an additional free variable can be introduced. One effective approach is to apply distinct bias voltages to the gates of M_0 and M_1, introducing a new variable, V_{b1}. As a result, the previously defined constraint gains additional degrees of freedom, enabling further optimization. This modified structure also reduces the voltage drop across the signal envelope (drain-source voltage), leading to lower power dissipation. The core schematic of this improved circuit is illustrated in Fig. 5.2 (right circuit), where separate bias voltages (V_b) are assigned to each transistor gate. The drain current of $M_{0,1}$ can be expressed through equations by incorporating $V_{b0,1}$ as substitution parameters. When $V_{env} = V_H$ and $V_{env} = V_L$:

$$V_{SPH} \approx V_H - \frac{1}{2i_s}\frac{(V_{b0} - V_{thn})^2}{(V_H - V_{b1} - |V_{thp}|)}, \quad V_{SPL} \approx \frac{i_s}{2}\frac{(V_L - V_{b1} - |V_{thp}|)^2}{(V_{b0} - V_{thn})} \quad (5.4)$$

showing that by using different voltage biases, ΔV_{SP} can be precisely optimized to set V_{SPH} and V_{SPL} maximum and minimum, respectively. Since the saturated transistor determines the current, increasing V_{b1} and decreasing V_{b0} should reduce the current consumption in both cases.

The MATLAB simulation results for a modulation index of 0.05, with $i_s = 1.4$ and $V_H = 2.2\,V$, show that ΔV_{SP} increases when V_{b1} is larger while V_{b0} is simultaneously reduced. This observation confirms that employing separate bias voltages (V_{b0} and V_{b1}) enhances ΔV_{SP}, offering improved signal distinction. Further robustness simulations, incorporating boundary conditions for the separated bias voltages at a modulation index of $MI = 0.10$, demonstrate that the feasible operating region is defined by specific constraints on V_{b0} and V_{b1}. Under the condition $V_{b0} = V_{b1}$, the feasible region decreases as i_s increases. To optimize robustness, the solution can be shifted towards an unbalanced voltage bias configuration, where $V_{b0} \neq V_{b1}$. Additionally, reducing i_s also enhances robustness. Therefore, a trade-off is observed between maximizing ΔV_{SP} and ensuring robustness, requiring careful parameter selection based on system constraints.

The self-sampling ASK demodulator is improved using a separated biasing voltage method, resulting in a novel structure, as shown in Fig. 5.3. This design incorporates two distinct biasing voltages for the digital shaper section, referred to as a separated-biasing structure. The middle transistors (M_2 to M_7) are arranged in a positive feedback configuration, where the output signal is utilized as an enable signal to generate two stable, low-noise biasing voltages. Additionally, gate capacitors within the digital shaper section are charged to further stabilize these voltages. The bulk terminals of the PMOS transistors are connected to the highest voltage in the circuit via dynamic bulk biasing (DBB) blocks. The biasing voltage levels can be adjusted by modifying the width-to-length ratio of the biasing generator circuit. The transistor dimensions, including size and number of fingers, as well as capacitor values, are detailed in Table 5.1. The simulation results, presented in Fig. 5.4a, depict the input modulated voltage ($V_{RX+} - V_{RX-}$), input data, the envelope of the input voltage (V_{env}), the output of the digital shaper (V_{sp}), and the demodulated data. A startup period is required initially to achieve stable data demodulation, ensuring a steady V_{sp}. Figure 5.4b provides a breakdown of the power consumption based on simulation results. The Shmoo plot in Fig. 5.5a illustrates the demodulator's performance across different process corners,

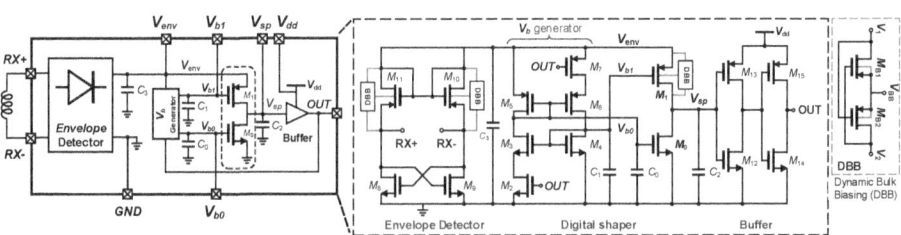

Fig. 5.3 Block diagram of the presented separated V_b ASK demodulator

5.3 Separated-Vb ASK Demodulator

Table 5.1 Transistors dimensions and capacitor values in Fig. 5.3

Device	#finger $\times (W/L)$ (nm)	Device	#fingers $\times (W/L)$ (nm)
M_0	$6 \times (240/680)$	$M_{10,11}$	$16 \times (1200/350)$
M_1	$14 \times (720/680)$	$M_{12,14}$	$1 \times (240/180)$
$M_{2,3,4}$	$1 \times (240/680)$	M_{13}	$1 \times (1440/180)$
M_5	$1 \times (720/680)$	M_{15}	$2 \times (1440/180)$
$M_{6,7}$	$4 \times (720/680)$	$C_{0,1,2}$	(652.86 fF)
$M_{8,9}$	$8 \times (300/350)$	C_3	(8.96 pF)

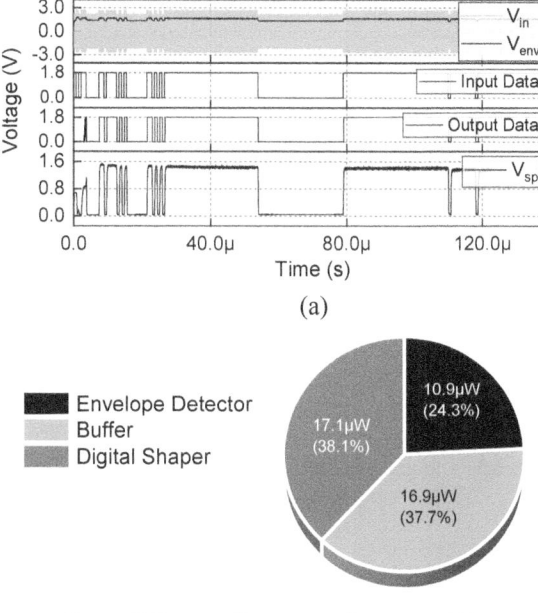

Fig. 5.4 a Post-layout simulation results of the demodulator, **b** breakdown of the simulated power dissipation

temperatures, and modulation indices. These results validate the demodulator's reliability for biomedical implant applications, even at modulation indices as low as 10%. Additionally, Fig. 5.5b, c show the stability analysis of the generated biasing voltages, V_{b0} and V_{b1}, respectively.

The valid range of V_L is a critical parameter for this design. In this implementation, the buffer is powered by a 1.8 V supply, while the input signal to the digital shaper, V_{env}, may exceed this voltage level. Consequently, if the output of the digital shaper surpasses the buffer's threshold voltage (approximately 0.9 V) when $V_{RX} = V_{RX+} - V_{RX-}$ is at a low

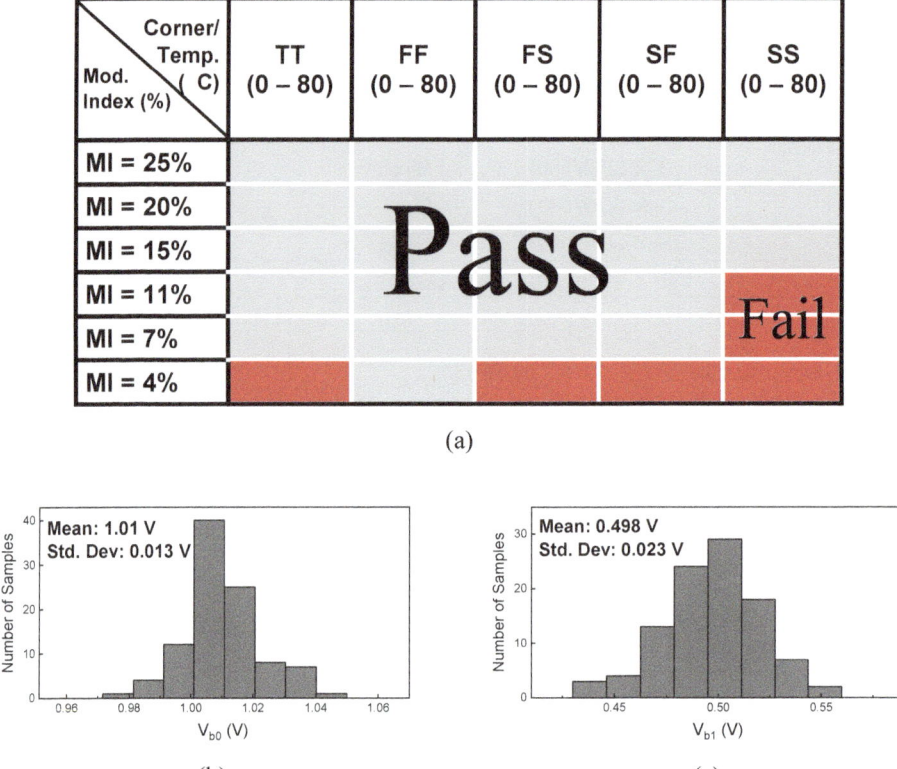

Fig. 5.5 Simulation results of **a** Shmoo plot illustrating various modulation indexes versus temperature variation across process corners, **b** V_{b0}, and **c** V_{b1}

level, the buffer output will be '1'. Additionally, V_{RX} must be maintained above a certain threshold to ensure sufficient power delivery to the implant, which is set at 2.2 V. Simulation results indicate that the valid range for the low-level voltage, $V_{RX} = V_{RXL}$, extends from 2 to 3.5 V when the modulation index is 0.1. The lower boundary of this range can be further reduced if a higher modulation index is allowed. The observed delay between the transmitted and demodulated data is approximately 200 ns for a frequency of 13.56 MHz. The data rate can be configured such that each bit is transmitted using 15–20 cycles of the carrier signal. Since the data rate is closely linked to signal delay, the delay between the envelope signal and the demodulated digital signal remains around 200 ns, enabling a theoretical data rate of approximately 5 Mbps. However, simulations reveal that if fewer than five carrier cycles are used per bit, output data distortion occurs, reducing the achievable data rate to 2.7 Mbps at 13.56 MHz. The power consumption breakdown of the demodulator components is as follows: the envelope detector consumes 10.9 μW, the digital shaper 17.1 μW, and the buffer

5.3 Separated-Vb ASK Demodulator

16.9 µW, leading to a total power consumption of 45 µW. The average simulated current consumption is approximately 33.8 µA when V_{RXL} is 2.2 V.

5.3.2 Measurement Results

The presented separated-V_b ASK demodulator is designed and fabricated using a 180 nm standard CMOS technology with an area of 0.00967 mm^2, as shown in Fig. 5.6. The layout has dimensions equal to 137.89 µm × 70 µm. The largest capacitor, C_3, functions as the rectifier capacitor in the envelope detector. The remaining smaller capacitors (C_0, C_1, and C_2) are used to stabilize the biasing voltage and filter the digital shaper's output signal. The experimental characterization is conducted by varying key parameters such as data and carrier frequencies, MI, and input voltage. The measured data are extracted from the oscilloscope for analysis. A 2^7-1 pseudo-random data sequence with a sampling rate of 100 Hz is generated and modulated onto a 10–20 MHz carrier signal (e.g., 13.56 MHz) using a signal generator. This modulated signal is then applied to the input of the demodulator. The measurement results, shown in Fig. 5.7, illustrate the input signal, along with the corresponding demodulated data, V_{sp}, and V_{env}. The total current consumption of the demodulator is measured at 25.4 µA. The performance of the proposed design across different data rates is presented in Fig. 5.8. Experimental results indicate that the demodulator operates with a minimum modulation index of approximately 0.08. To assess the maximum data rate capability, a pulse-based pseudo-random signal with a frequency range of 1–1.5 MHz (equivalent to 2–3 Mbps) is applied to the demodulator. The maximum observed operating data rate for demodulation is 2.7 Mbps. To evaluate the bit error rate (BER), a 2^7-1 pseudo-random binary sequence (PRBS7) is generated at varying data rates (ranging from 100 kbps to 2.5 Mbps) and applied to the demodulator input. No errors were detected in data sequences exceeding

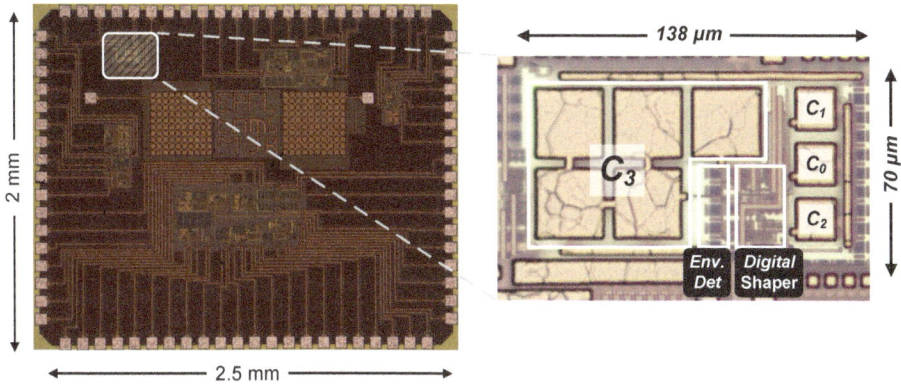

Fig. 5.6 Die photograph of the separated-V_b ASK demodulator

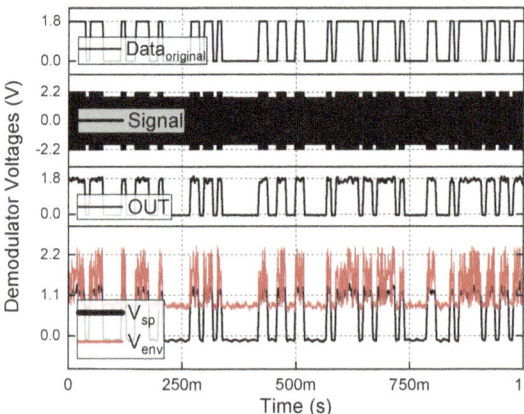

Fig. 5.7 Measurement results of pseudo-random data demodulation, the digital shaper voltage V_{sp}, and the envelope voltage V_{env}

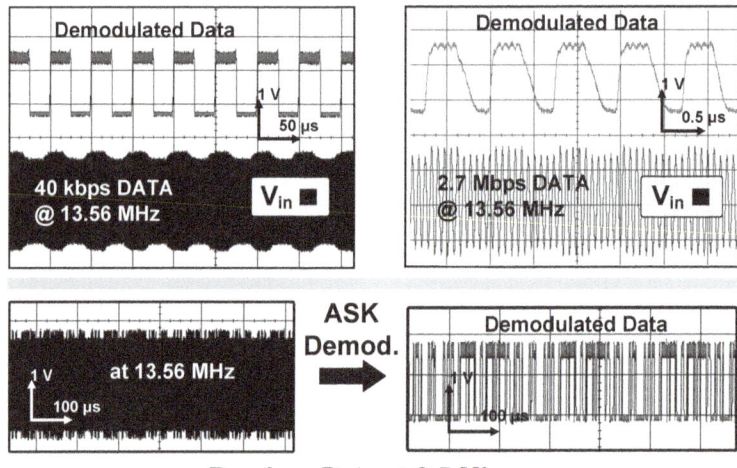

Fig. 5.8 Measured results of demodulated data in different data rates at 10% MI

one million bits at 0.1 Mbps, yielding a BER below 10^{-6}. At higher data rates, the BER was measured as 8.6×10^{-3} at 0.5 Mbps and 3.1×10^{-2} at 2.5 Mbps.

A figure of merit (FoM) is defined to evaluate the performance of ASK demodulators, as expressed in Eq. (5.5). This metric considers energy consumption per bit, the relationship between circuit area and CMOS technology, and the modulation index, providing a comprehensive assessment of the demodulator's efficiency.

$$\text{FoM} = \frac{\text{Data Rate}}{\text{Power}} \times \frac{1}{\text{Area/Technology}} \times \frac{1}{MI} \tag{5.5}$$

5.4 Averaging ASK Demodulator 113

The presented ASK demodulator offers a relatively significant advantage in FoM, attributed to its high data rate, low power consumption, and compact chip area. A comparative analysis of the proposed ASK demodulator with respect to state-of-the-art designs, along with their respective FoM values, is presented in Table 5.2.

5.4 Averaging ASK Demodulator

An ASK demodulator specifically designed for uplink LSK demodulation is illustrated in Fig. 5.9. This demodulator enhances data rates and includes three components: an envelope detector, a buffer with an averaging circuit, and a comparator. The envelope detector serves the purpose of extracting the data embedded within the sine wave envelope, subsequently transmitting it to a buffer and low-pass filter for data recovery. Additionally, it extracts the DC voltage of the envelope and feeds it to the envelope detector [18] through the utilization of off-chip components, namely large resistors and capacitors [5].

5.4.1 Circuit Architecture

In the envelope detector shown in Fig. 5.9, initially, a fixed current source charges node V_{env}. Whenever the input voltage changes, the drain currents of M_3 and M_4 become imbalanced, resulting in the creation of a current pulse on M_5 [19]. This current pulse discharges the output node through the current mirror consisting of M_5 and M_6, causing it to follow the transition of V_{in}. As V_{env} approaches V_{in}, the difference in current between M_1 and M_2 gradually approaches zero. A low-pass filter is employed to differentiate between the high and low output voltage states of the envelope detector and to effectively demodulate data. A sufficiently large time constant must be chosen with a passive RC low-pass filter. Consequently, a pseudo-resistor is introduced using a PMOS transistor. This pseudo-resistor effectively simulates the behavior of a TΩ resistor. The recovery and demodulation of uplink data involve a comparison with the average voltage of converted voltages. This operation is executed by employing a comparator in the final stage of the demodulator. The output of this comparator provides the signal at its full amplitude. Figure 5.10a shows the simulation results of the designed averaging ASK demodulator, including the input voltage, demodulated data, envelope, and average voltage.

5.4.2 Measurement Results

The demodulation circuit has been developed and fabricated using 180 nm CMOS technology. The chip photograph is presented in Fig. 5.11. The experimental results of the demodulator are shown in Fig. 5.10b. The minimum modulation index for the demodula-

Table 5.2 ASK demodulator benchmarking [13]

Paper	Source	Year	$f_{carrier}$ (MHz)	V_{dd} (V)	Data rate (Mbps)	Area (μm^2)	CMOS (μm)	Power (μW)	MI (min) (%)	Energy (pJ/bit)	S/M*	FoM
[14]	TCAS II	2015	13.56	1.8	6.78	80000	0.18	350	2.56%	51.62	M	5.95
[15]	BioCAS	2016	13.56	3	0.7	14774	0.35	76.5	1.10%	109.29	S	15.07
[16]	IET Journal	2016	13.56	1.8	2	3000	0.18	35	7%	17.5	M	17.14
[10]	TCAS II	2017	5	1.8	0.5	920	0.18	17	5%	34	M	19.56
[17]	JSSC	2020	13.56	1.8	0.1	–	0.065	132	0.10%	1320	M	–
[5]	TBCAS	2022	13.56	1.8	1	514800	0.18	280	10%	280	M	0.034
This work		2023	13.56	1.8	2.7	1440	0.18	45	10%	16.67	M	33.75

*S: simulation, M: measurement

5.5 FSK Demodulator

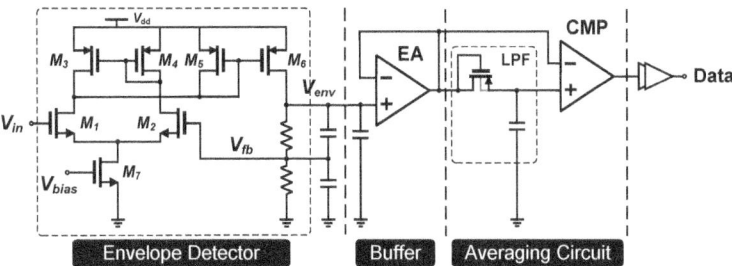

Fig. 5.9 Schematic of the averaging ASK demodulator

tor is 24%, with a maximum data rate of 1.1 Mbps, operating at the carrier frequency of 13.56 MHz.

5.5 FSK Demodulator

In the downlink data path from the external unit to the CIU [20], frequency shift keying (FSK) is employed to facilitate data reception without interfering with power transmission. Additionally, frequency modulation exhibits greater resilience to noise, enhancing robustness and reliability in terms of signal-to-noise ratio and overall signal quality compared to other modulation techniques. The schematic of the FSK demodulator is depicted in Fig. 5.12. The design consists of three key components: a clock recovery circuit, a frequency-to-amplitude converter (F2V), and a CMOS current reference.

5.5.1 Circuit Architecture

The circuit diagram of the proposed FSK demodulator is shown in Fig. 5.13. The clock regeneration is required to convert the FSK-modulated sine wave signal into a digital pulse, employing an inverter with hysteresis behavior. Two connected cross-coupled inverters are designed for the hysteresis behavior. The F2V conversion circuit consists of a current source, a capacitor, and a switch [21]. When the recovered clock signal is low, the capacitor is connected to the current source and gradually charges. Upon transitioning to a high state, the switch connects the capacitor to the ground, causing it to discharge. Since the capacitor linearly charges over time, the resulting output voltage is directly proportional to the duration and period of the clock's low state. A sample-and-hold circuit stores this voltage just before reset, effectively converting the frequency or period of the clock into a corresponding voltage level. To distinguish between the high and low states of the F2V converter's output and correctly demodulate data, a low-pass filter is employed to establish an average voltage that represents the midpoint between the two states. However, to prevent modulation artifacts

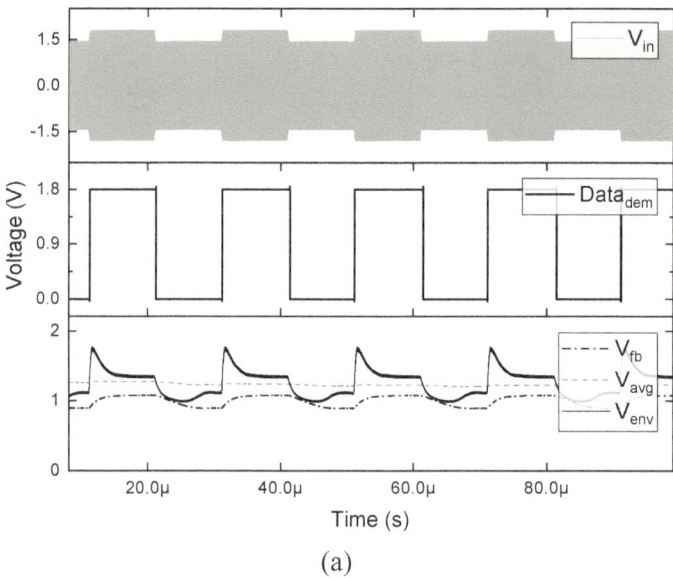

(a)

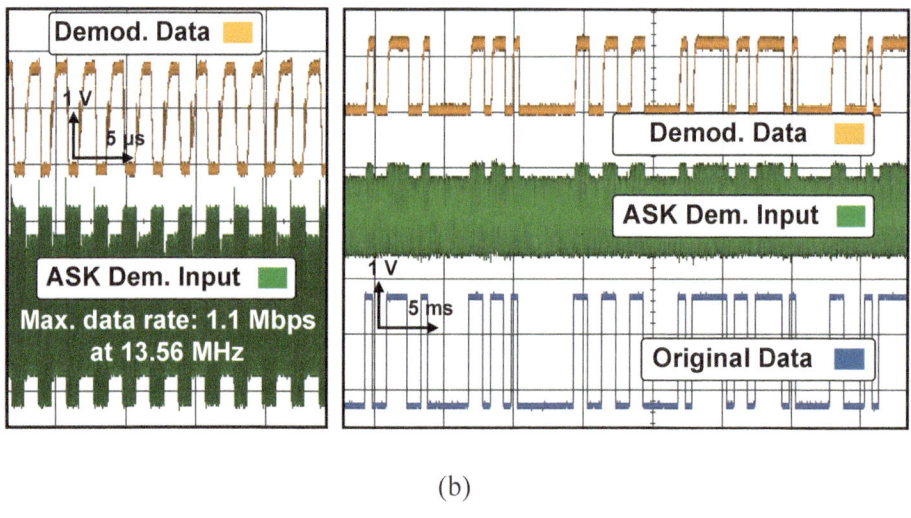

(b)

Fig. 5.10 **a** Simulation and **b** measurement results of the averaging ASK data demodulation

5.5 FSK Demodulator

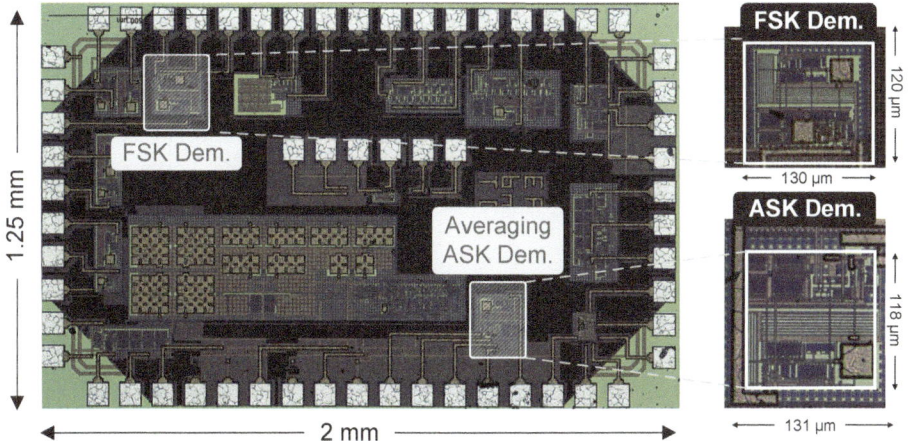

Fig. 5.11 Die photograph of the FSK and averaging ASK demodulator

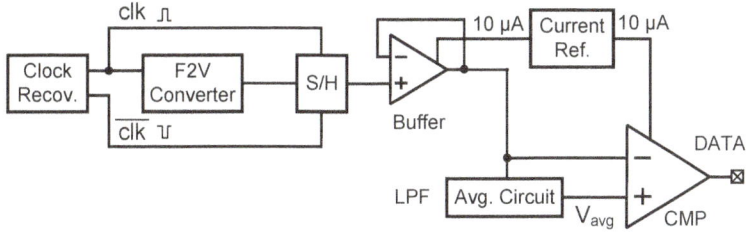

Fig. 5.12 Block diagram of the proposed FSK demodulator

from appearing at the output of the low-pass filter, the circuit must maintain a sufficiently large time constant. This requirement is addressed through the design of a passive RC low-pass filter. Ensuring a sufficiently large time constant requires either a capacitor in the μF range or a resistor in the $T\Omega$ range. However, integrating such a large capacitor on-chip is impractical due to excessive area consumption. To overcome this limitation, a PMOS transistor is implemented as a pseudo-resistor. At the final stage, a comparator is used to demodulate and recover the downlink data by comparing the converted signal with the average voltage. The comparator's output provides a fully restored digital signal. The simulation results of the proposed FSK demodulator circuit are shown in Fig. 5.14a, depicting the input voltage, recovered clock, original and demodulated data, and the output of the low-pass filter.

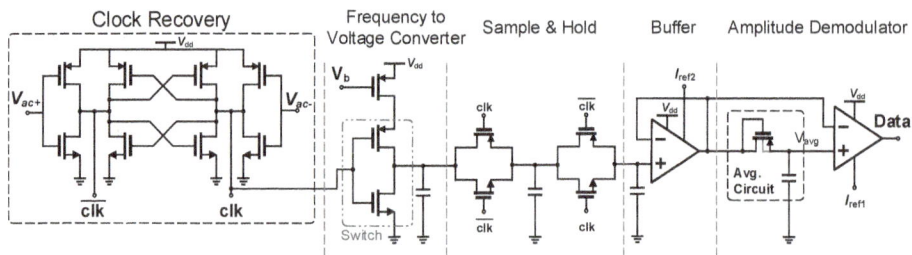

Fig. 5.13 Schematic of the data conversion unit including frequency shift keying (FSK) demodulator, frequency to voltage (F2V) converter, CMOS current reference, and clock recovery

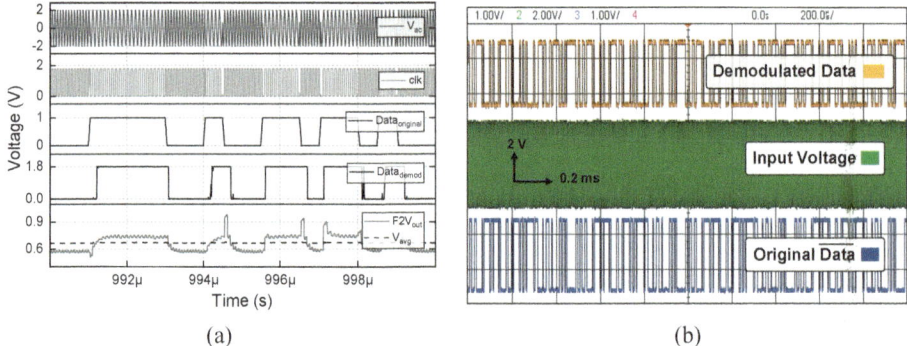

Fig. 5.14 a Simulation and **b** measurement results of the FSK data demodulation

5.5.2 Measurement Results

The presented FSK demodulator has been developed and fabricated in a 180 nm CMOS technology, as shown in Fig. 5.11. To evaluate the performance of the demodulator, a 100-bit pseudo-random data stream with a sampling rate of 100 Hz is generated and modulated onto a 13.56 MHz carrier using a signal generator. This modulated signal is then applied to the demodulator as the received downlink data. The input voltage of the demodulator, the original data, and the demodulated output are illustrated in Fig. 5.14b. It is important to note that the demodulator requires a minimum of 10 carrier cycles to accurately convert data. As a result, the maximum achievable data rate is 1.3 Mbps. Additionally, the minimum operating modulation index, defined as $(f_1 - f_2)/f_1$, is 10%. The total current consumption of the data conversion unit is measured at 0.385 mA when operating at 1.8 V. The benchmarking and performance of the proposed FSK demodulator are shown in Table 5.3. According to the comparison table, a trade-off emerges between power consumption and modulation index. The demodulation process relies on frequency-to-voltage conversion, which is in the analog domain. To recover the FSK data modulated with a low modulation index, however, data demodulation in the digital domain can be used.

Table 5.3 Benchmarking of the proposed FSK demodulator and previous works [20]

	This work	[22]	[21]	[23]	[24]	[25]
Carrier frequency	13.56 MHz	200–333.3 KHz	5 MHz	13–16 MHz	6.78, 13.56 MHz	6.5–7.5 MHz
Demodulation type	Analog	Digital	Analog	Analog/digital	PLL based	Analog
Max. data rate	1.3 Mbps	133 Kbps	5 Mbps	1 Mbps	1 Mbps	2.5 Mbps
Modulation index	10%	66%	1%	23%	–	15%
Area (mm^2)	0.015	–	0.22	0.023	0.084	–
Technology (nm)	180	180	180	180	350	180
Supply voltage (V)	1.8	1.8	1.8	1	1.5	4
Power (mW)	0.69	0.034	1	0.042	0.084	0.24

5.6 Summary

The design and characterization of the data communication unit are presented in this Chapter. Different modulation techniques are proposed with their circuit architecture that achieve robust and low-power data and clock recovery. These designed blocks are measured with a single inductive link in the CIU and ASPs. Additionally, this Chapter presents the required circuits for bidirectional telemetry, including:

- A clock recovery.
- A LSK demodulator.
- A separated-V_b ASK demodulator for the ASPs which offers low-power and high data rate demodulation by introducing a novel circuit improvement to provide larger circuit design flexibility and delay improvement.
- An averaging demodulator for uplink data communication within the CIU, to demodulate data transmitted from the ASPs to the CIU.
- A FSK demodulator for the data communication between the external unit and the CIU.

The ASIC was designed using a standard 180 nm CMOS process. The performance of the data units is verified with separate on-chip measurements.

References

1. M. J. Karimi, M. Jin, Y. Zhou, C. Dehollain, and A. Schmid, "Wirelessly Powered and Bi-directional Data Communication System with Adaptive Conversion Chain for Multisite Biomedical Implants Over Single Inductive Link," *IEEE Transactions on Biomedical Circuits and Systems*, pp. 1–11, 2024.
2. M. Lotfi Navaii, H. Sadjedi, and A. Sarrafzadeh, "Efficient ASK Data and Power Transmission by the Class-E With a Switchable Tuned Network," *IEEE Transactions on Circuits and Systems I: Regular Papers*, vol. 65, no. 10, pp. 3255–3266, 2018.
3. B. Lee, M. Kiani, and M. Ghovanloo, "A Smart Wirelessly Powered Homecage for Long-Term High-Throughput Behavioral Experiments," *IEEE Sensors Journal*, vol. 15, pp. 4905–4916, 2015.
4. D. Jiang, D. Cirmirakis, M. Schormans, T. A. Perkins, N. Donaldson, and A. Demosthenous, "An Integrated Passive Phase-Shift Keying Modulator for Biomedical Implants With Power Telemetry Over a Single Inductive Link," *IEEE Transactions on Biomedical Circuits and Systems*, vol. 11, no. 1, pp. 64–77, 2017.
5. Y. Chen, Y. Liu, Y. Li, G. Wang, and M. Chen, "An Energy-Efficient ASK Demodulator Robust to Power-Carrier-Interference for Inductive Power and Data Telemetry," *IEEE Transactions on Biomedical Circuits and Systems*, vol. 16, no. 1, pp. 108–118, 2022.
6. Q. Zhang, S. Mai, R. Zhou, and X. Yang, "A Low-power ASK Demodulator for Wireless Power and Data Transfer Systems Supporting Ultra-low Modulation Depth of 0.03%," in *2023 IEEE International Symposium on Circuits and Systems (ISCAS)*, pp. 1–5, 2023.
7. H. Yu and K. Najafi, "Low-power interface circuits for bio-implantable microsystems," in *2003 IEEE International Solid-State Circuits Conference, 2003. Digest of Technical Papers. ISSCC.*, pp. 194–487, IEEE, 2003.
8. C.-C. Wang, Y.-H. Hsueh, U. F. Chio, and Y.-T. Hsiao, "A C-less ASK demodulator for implantable neural interfacing chips," in *2004 IEEE International Symposium on Circuits and Systems (IEEE Cat. No. 04CH37512)*, vol. 4, pp. Iv–57, IEEE, 2004.
9. T.-J. Lee, C.-L. Lee, Y.-J. Ciou, C.-C. Huang, and C.-C. Wang, "C-less and R-less low-frequency ASK demodulator for wireless implantable devices," in *2007 International Symposium on Integrated Circuits*, pp. 604–607, IEEE, 2007.
10. M. L. Navaii, M. Jalali, and H. Sadjedi, "A 34-pJ/bit Area-Efficient ASK Demodulator Based on Switching-Mode Signal Shaping," *IEEE Transactions on Circuits and Systems II: Express Briefs*, vol. 64, pp. 640–644, 2017.
11. C. C. Wang, C. L. Chen, R. C. Kuo, and D. Shmilovitz, "Self-sampled all-MOS ASK demodulator for lower ISM band applications," *IEEE Transactions on Circuits and Systems II: Express Briefs*, vol. 57, pp. 265–269, 2010.
12. G. Yilmaz, O. Atasoy, and C. Dehollain, "Wireless energy and data transfer for in-vivo epileptic focus localization," *IEEE Sensors Journal*, vol. 13, pp. 4172–4179, 2013.
13. M. J. Karimi, Y. Zhou, C. Dehollain, and A. Schmid, "An Analysis of An ASK Demodulator With Dual Self-Biased Separated Voltages for Implantable Applications," *IEEE Transactions on Circuits and Systems II: Express Briefs*, pp. 1–1, 2024.
14. H. Lee, J. Kim, D. Ha, T. Kim, and S. Kim, "Differentiating ASK Demodulator for Contactless Smart Cards Supporting VHBR," *IEEE Transactions on Circuits and Systems II: Express Briefs*, vol. 62, no. 7, pp. 641–645, 2015.
15. H. Zhang, X. Chen, M. Chen, and G. Wang, "A wide-input-range low-power ASK demodulator for wireless data transmission in retinal prosthesis," in *2016 IEEE Biomedical Circuits and Systems Conference (BioCAS)*, pp. 492–495, 2016.

16. N. Mousavi, M. Sharifkhani, and M. Jalali, "Ultra-low power current mode all- MOS ASK demodulator for radio frequency identification applications," *IET Circuits, Devices & Systems*, vol. 10, no. 2, pp. 130–134, 2016.
17. D. Ye, Y. Wang, Y. Xiang, L. Lyu, H. Min, and C. J. Shi, "A Wireless Power and Data Transfer Receiver Achieving 75.4% Effective Power Conversion Efficiency and Supporting 0.1% Modulation Depth for ASK Demodulation," *IEEE Journal of Solid-State Circuits*, vol. 55, no. 5, pp. 1386–1400, 2020.
18. S. Sanielevici, K. Cioffi, B. Ahrari, P. Stephenson, D. Skoglund, and M. Zargari, "A 900-MHz transceiver chipset for two-way paging applications," *IEEE Journal of Solid-State Circuits*, vol. 33, no. 12, pp. 2160–2168, 1998.
19. M. J. Karimi, S. Mehdi, C. Dehollain, and A. Schmid, "Wireless Power and Data Transceiver in A Central Implanted Unit for Biomedical Applications," in *2024 IEEE 15th Latin America Symposium on Circuits and Systems (LASCAS)*, pp. 1–5, 2024.
20. M. J. Karimi, S. Mehdi, C. Dehollain, and A. Schmid, "A Wireless Power and Bi-Directional Data Transfer System Using A Single Inductive Link for Biomedical Implants," in *2023 IEEE Biomedical Circuits and Systems Conference (BioCAS)*, pp. 1–5, 2023.
21. A. A. Razavi Haeri and A. Safarian, "A Cycle by Cycle FSK Demodulator With High Sensitivity of 1% Frequency Modulation Index for Implantable Medical Devices," *IEEE Transactions on Circuits and Systems I: Regular Papers*, vol. 69, no. 11, pp. 4682–4690, 2022.
22. C. H. Kao, Y. P. Lin, and K. T. Tang, "Wireless data and power transmission circuits in biomedical implantable applications," *Proceedings of 2011 International Symposium on Bioelectronics and Bioinformatics*, ISBB 2011, pp. 9–12, 2011.
23. Y.-T. Hou and P.-H. Hsieh, "A 1-Mbps Frequency-Shift Keying Receiver for Inductively Powered Biomedical Applications," in *2020 IEEE Asia-Pacific Microwave Conference (APMC)*, pp. 491–492, 2020.
24. Y.-S. Hwang, B.-H. Hwang, H.-C. Lin, and J.-J. Chen, "PLL-Based Contactless Energy Transfer Analog FSK Demodulator Using High-Efficiency Rectifier," *IEEE Transactions on Industrial Electronics*, vol. 60, no. 1, pp. 280–290, 2013.
25. Y. Park, S. T. Koh, J. Lee, H. Kim, J. Choi, S. Ha, C. Kim, and M. Je, "A Wireless Power and Data Transfer IC for Neural Prostheses Using a Single Inductive Link with Frequency-Splitting Characteristic," *IEEE Transactions on Biomedical Circuits and Systems*, vol. 15, no. 6, pp. 1306–1319, 2021.

Towards On-Chip CMOS Temperature Sensing

Temperature sensing is an important part of implanted biomedical devices. For instance, temperature sensing can be used in various biomedical applications such as health monitoring or smart implants. The accurate measurement of temperature is critical for these applications to detect early signs of diseases or device overheating. It also enables the monitoring of the health status of a patient. Temperature sensing is also required to prevent excessive wireless powering and control thermal requirements inside the body or brain [1, 2]. Also, to minimize the device area, the inclusion of an on-chip temperature sensor is essential. To achieve these goals, temperature sensors can be integrated into each biomedical implant or smart tag, as shown in Fig. 6.1. Hence, the development of ultra low-power and high-precision temperature sensors has become an active research area in recent years [3–5].

A temperature sensor for biomedical applications should exhibit energy and area efficiency and provide maximum temperature sensitivity. Temperature is mainly converted into an analog electrical signal, such as voltage, current, delay, and frequency. Subsequently, this signal can be translated into a digital signal to support thermal management. On-chip temperature sensors are used to measure temperature using a variety of devices, such as resistors [6], BJTs [3, 4, 7], MOS gate leakage [8] and MOS transistors [9–12]. BJT-based sensors use bipolar proportional to the absolute temperature (PTAT) voltage with a $\Sigma\Delta$-ADC [13], while resistor-based sensors utilize the voltage variation sensed by resistors [6]. A SAR-based temperature sensor is proposed in [14] that uses an unbalanced structure in the comparator. MOS transistor-based sensors employ proportional and complementary-to-absolute-temperature (PTAT and CTAT) voltage generation by biasing and sizing transistors in the sub-threshold region. This approach results in low power consumption. Signal amplification or integration can also be used to enhance temperature sensitivity and accuracy at the cost of energy and area consumption. This chapter presents a fully integrated CMOS temperature sensor. It utilizes CTAT/PTAT voltage comparison and a sensitivity improve-

© The Author(s), under exclusive license to Springer Nature Switzerland AG 2026
M. J. Karimi et al., *Integrated Wireless Power, Data Communication, and Thermal Sensing Systems for Autonomous Multisite Brain Implants*, Synthesis Lectures on Engineering, Science, and Technology, https://doi.org/10.1007/978-3-031-90839-2_6

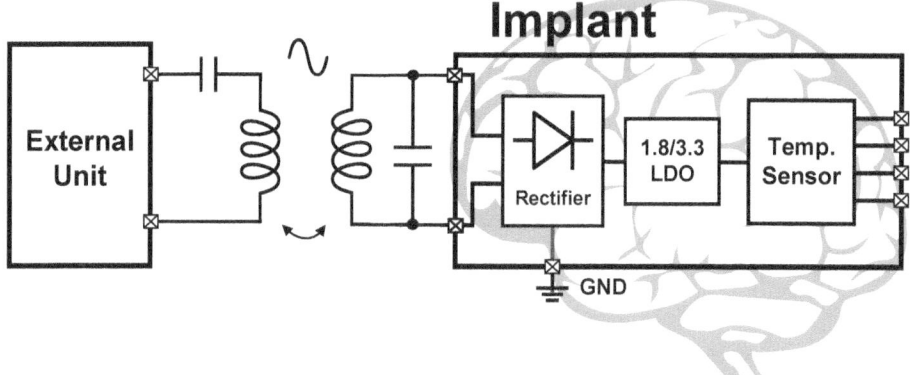

Fig. 6.1 Overview of the integrated temperature sensor in the biomedical implants

ment circuit. The proposed temperature sensor is designed to operate in the subthreshold region.

6.1 CMOS Sensor Architecture

The block diagram of the proposed temperature sensor is shown in Fig. 6.2. Temperature is converted to PTAT and CTAT voltages via core circuits. The core voltages are then subtracted to enhance the sensitivity and voltage-to-temperature (VTT) slope. The output voltage is fed to the voltage-to-current converter (VIC) block. It employs on-chip negative and positive temperature coefficient resistors (NTC and PTC) to further enhance the sensitivity and VTT slope of the sensor. Finally, this voltage feeds the input voltage of the designed 10-bit successive approximation register (SAR) analog-to-digital converter (ADC) via an intermediary buffer for proper input driving. Fig. 6.3 shows the schematic of the temperature sensor core, which comprises PTAT and CTAT cores and a current reference. The startup circuit initializes the subthreshold current source in the core, which generates currents that feed the PTAT and CTAT voltage generations with specific ratios. The PTAT and CTAT behavior is determined by the ratio between the dimensions of M_1 and M_2 (in Fig. 6.4) for each core subcircuit, as per the circuit calculations. To enhance the voltage differences throughout the temperature sweep, high and low voltage threshold transistors are employed in the core. This is achieved by adjusting the dimensions of the transistors to preserve the linearity of the VTT slope. For an MOS transistor biased in the sub-threshold region:

$$I_{\text{sub-th}} = I_0 \left(\frac{W}{L}\right) \exp\left(\frac{V_{GS} - V_{th}}{\eta V_T}\right) \cdot \left[1 - \exp\left(-\frac{V_{DS}}{V_T}\right)\right] \approx I_0 \left(\frac{W}{L}\right) \exp\left(\frac{V_{GS} - V_{th}}{\eta V_T}\right) \tag{6.1}$$

6.1 CMOS Sensor Architecture

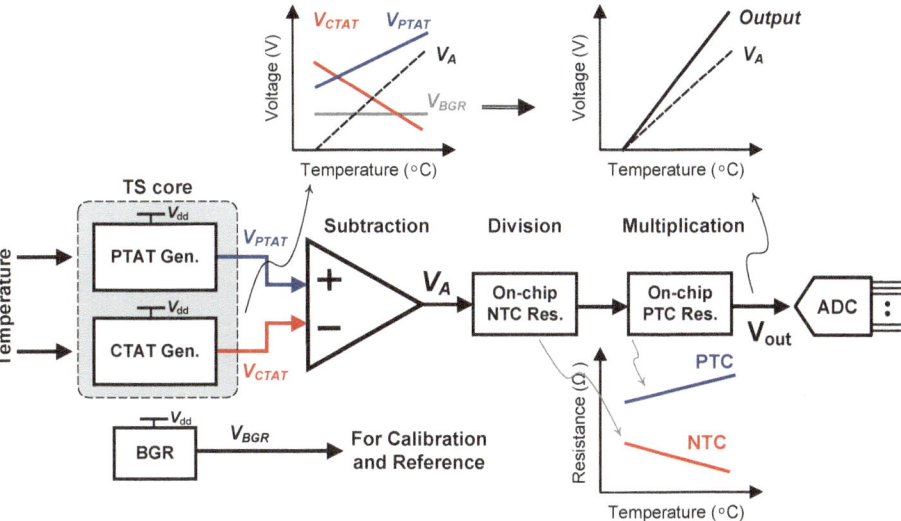

Fig. 6.2 Block diagram of the proposed temperature sensor

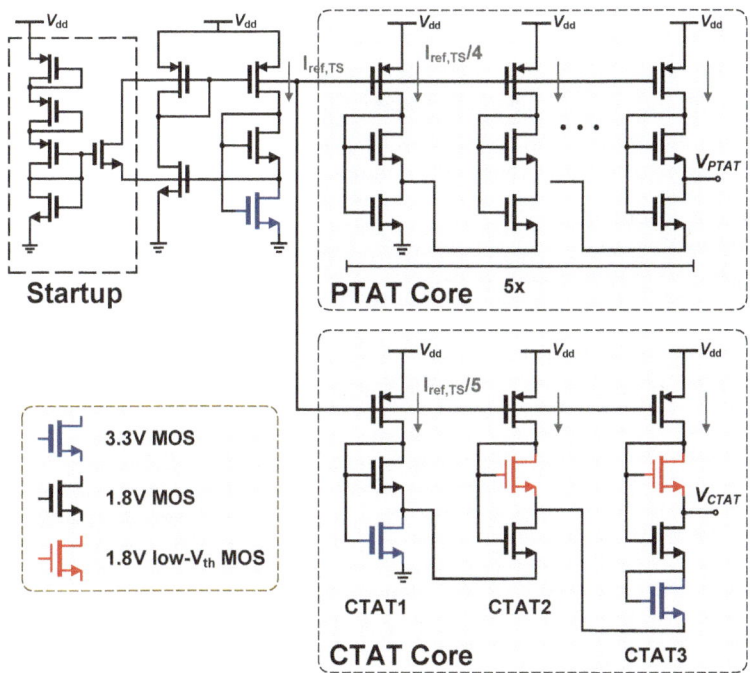

Fig. 6.3 Schematic of the temperature sensor (TS) core, including the PTAT and CTAT voltage generator and current reference

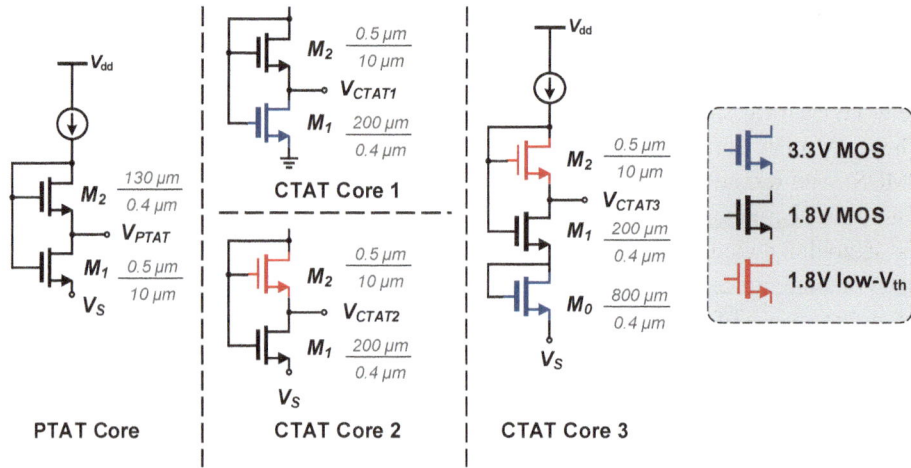

Fig. 6.4 Schematic of PTAT and CTAT cores with the transistor sizing

$$V_{GS} = \eta V_T \ln\left(\frac{I_{\text{sub-th}}}{I_0 \cdot (W/L)}\right) + V_{th} \qquad (6.2)$$

where I_0 is a process-dependent constant; and W, L, V_{GS}, V_{DS}, V_{th}, η, and V_T are the transistor width and length, gate-source voltage, drain-source voltage, voltage threshold, the inclination of the curve in weak inversion, and thermal voltage, respectively. Also, $V_T = kT/q$, where k is Boltzmann's constant, q is the elementary charge, and T is the temperature. Consequently, to achieve PTAT voltage with the positive temperature coefficient (TC), as shown in Fig. 6.4,

$$V_{\text{PTAT}} = V_{GS,M1} - V_{GS,M2}, \; \left(\frac{W}{L}\right)_{M1} < \left(\frac{W}{L}\right)_{M2} \Rightarrow \text{PTAT voltage} \qquad (6.3)$$

and for n stages of the PTAT core, the following equation is derived:

$$\Rightarrow V_{\text{PTAT}} = V_T \ln \frac{\left(\frac{W}{L}\right)_{M2}^n}{\left(\frac{W}{L}\right)_{M1}^n} = nV_T \underbrace{\frac{\left(\frac{W}{L}\right)_{M2}}{\left(\frac{W}{L}\right)_{M1}}}_{\text{positive +}} \qquad (6.4)$$

Furthermore, to design the CTAT core with the negative TC, the (W/L) of M_1 should be larger than M_2. Three CTAT core circuits are designed and combined to add the final CTAT voltage with the following equation:

$$\left(\frac{W}{L}\right)_{M1} > \left(\frac{W}{L}\right)_{M2} \Rightarrow \text{CTAT voltage} \qquad (6.5)$$

6.1 CMOS Sensor Architecture

$$V_{CTAT1} = V_T \ln \frac{\left(\frac{W}{L}\right)_{M2}}{\left(\frac{W}{L}\right)_{M1}} + \Delta V_{th,HVt\text{-}nom} \tag{6.6}$$

$$V_{CTAT2} = V_T \ln \frac{\left(\frac{W}{L}\right)_{M2}}{\left(\frac{W}{L}\right)_{M1}} + \Delta V_{th,nom\text{-}LVt} \tag{6.7}$$

$$V_{CTAT3} = V_T \ln \frac{\left(\frac{W}{L}\right)_{M2}}{\left(\frac{W}{L}\right)_{M1}} + \Delta V_{th,nom\text{-}LVt} + V_{th,HVt} \tag{6.8}$$

where $V_{th,HVt}$, $V_{th,nom}$, and $V_{th,LVt}$ are the voltage threshold of a 3.3 V, nominal 1.8 V, and low-V_{th} transistors, respectively, and $\Delta V_{th,HVt\text{-}nom}$ and $\Delta V_{th,nom\text{-}LVt}$ represent the differences in voltage thresholds. Specifically, $\Delta V_{th,HVt\text{-}nom}$ represents the difference between the voltage thresholds of 3.3 V high-V_{th} and nominal 1.8 V transistors, while $\Delta V_{th,nom\text{-}LVt}$ refers to the difference between nominal 1.8 V and low-V_{th} transistors. The simulation results of the voltage thresholds and their differences through temperature sweeps are shown in Fig. 6.5a. This figure illustrates the absence of variation in the voltage threshold difference when subjected to temperature sweeping. Moreover, Fig. 6.5b shows the output voltages of the PTAT and CTAT cores.

The schematic of the proposed VIC circuit for VTT slope enhancement is shown in Fig. 6.6. It consists of two amplifiers serving as comparators, negative and positive TC (NTC and PTC) on-chip resistors, and a bandgap reference (BGR) serving as a calibration reference with a zero temperature coefficient. The NTC resistor is a polysilicon resistor, whereas the PTC resistor is a p+ diffused resistor. On-chip resistors are subject to process-voltage-temperature (PVT) variations; however, the ratio between the two resistors remains unaffected by these changes. Consequently, the output voltage is dependent on the ratio of these two resistors, a principle that is reported in the design of bandgap references [15].

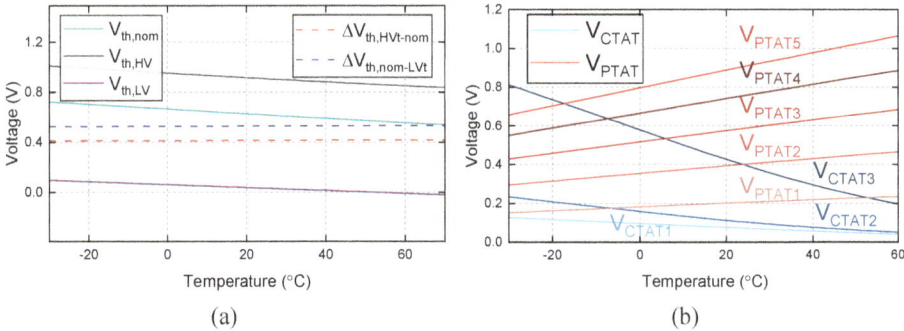

Fig. 6.5 Simulation results of the **a** voltage threshold and ΔV_{th} variations and **b** PTAT and CTAT core voltages through temperature sweep

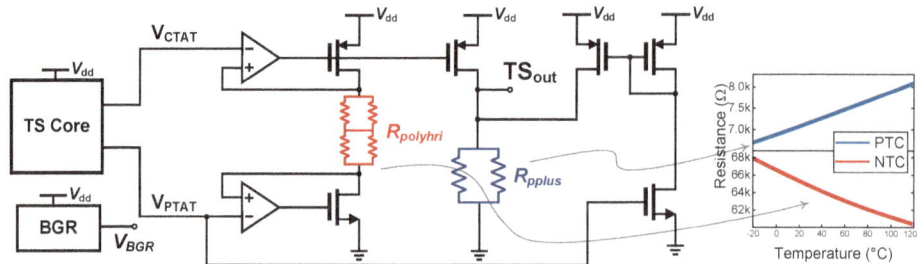

Fig. 6.6 Detailed schematic of the temperature sensor and voltage (V) to current (I) converter (VIC)

6.1.1 Bandgap Voltage Reference

The schematic of the BGR is shown in Fig. 6.7, which provides a reference voltage (0.49 V) and current (10 µA and 100 µA) with a very low temperature coefficient. The fundamental operational principle of bandgap references involves using bipolar transistors. This approach aims to achieve an almost zero temperature coefficient by summing two temperature coefficient terms—one negative and one positive. The presented BGR uses the low-voltage structure by employing R_X and R_Y ($R_Y = R_X$) [15]. According to the circuit schematic for the voltage of R_1:

$$|I_{D1}| = |I_{D2}| = \frac{V_T \ln n}{R_1} + \frac{|V_{BE1}|}{R_Y} \Rightarrow V_{BGR} = \frac{R_4}{R_X}\left(\frac{R_X}{R_1} V_T \ln n + |V_{BE1}|\right) \quad (6.9)$$

where $I_{D1,2}$ are the currents flowing through the current mirrors, V_{BE1} is the base-emitter voltage of the bipolar transistor, and V_T is the thermal voltage. This BGR offers a line

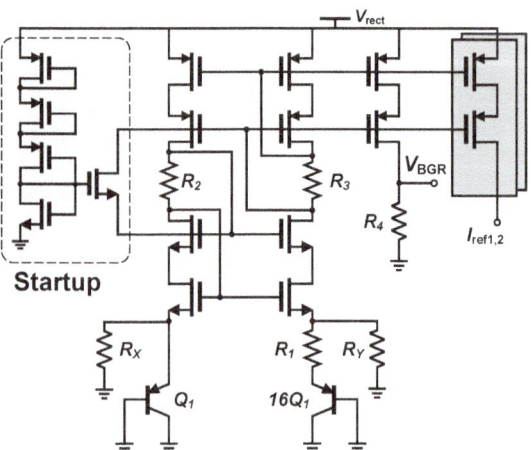

Fig. 6.7 Schematic of the bandgap voltage reference with its startup circuit

6.2 10-Bit SAR ADC Design

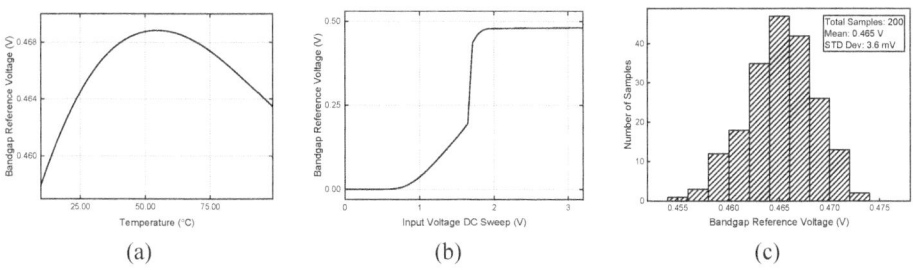

Fig. 6.8 Simulation results of the reference voltage through **a** temperature sweep, **b** DC sweep, and **c** Monte Carlo simulation

sensitivity of 6.27 mV/V. The temperature, DC sweep, and Monte Carlo simulation results of the voltage reference are shown in Fig. 6.8.

6.2 10-Bit SAR ADC Design

The schematic of the 10-bit SAR ADC used for the temperature sensor is shown in Fig. 6.9. It includes the sample and hold (S/H) block, SAR logic, clocked comparator, and digital-to-analog converter (DAC). The designed DAC utilizes the split binary weighted capacitive DAC. The operation of the ADC can be divided into three stages: (1) sampling, (2) conversion, and (3) recycling. In the sampling stage, the bottom plates of all the capacitors are connected to the ground while the upper plates are connected to the inverting terminal of the comparator. The non-inverting input of the comparator is connected to the sampled output. The second and third stages involve converting the sampled signal into digital form based on

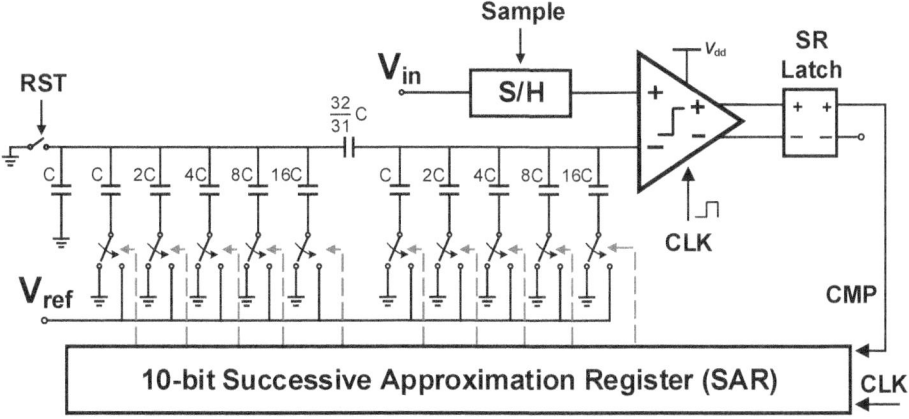

Fig. 6.9 Schematic of the 10-bit SAR ADC

Fig. 6.10 Schematic of the StrongARM comparator used in the 10-bit SAR ADC

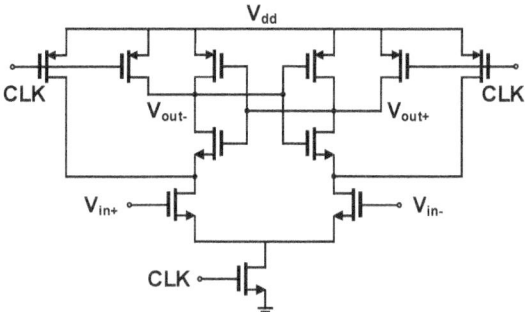

the comparator output and control signals of switches. The input signal is sampled using a sample-and-hold circuit. A sample-and-hold circuit is designed using a CMOS transmission gate and a capacitor. Figure 6.10 shows the designed StrongARM comparator adopted from [16] used for the 10-bit SAR ADC. It should be noted that a buffer is implemented between the temperature sensor and the ADC. This buffer is used to ensure proper driving of the ADC input. The operation of SAR ADC at high temperatures is studied in [17], demonstrating a variation of 0.3 LSB at 175 °C. It indicates that the design is suitable for our intended biomedical application within the temperature range of −30 to 110 °C.

6.3 Sensor Simulation Results

The results of DC-sweep simulations are presented in Fig. 6.11, showing the core output voltages with the VIC voltage. It also shows the results of a Monte Carlo simulation assessing the variability of the VIC voltage across various process variations. Figure 6.12 shows the simulated PTAT, CTAT, and VIC voltages throughout a temperature sweep. This figure also shows the Monte Carlo simulation results of the VIC and bandgap voltages. The voltage generated by the sensor subsequently serves as the input signal for the SAR ADC. The characteristics of the digital signal during the temperature sweep are presented at the bottom of Fig. 6.12. At room temperature, the sensor core exhibits a power consumption of 22.3 nW while the ADC dissipates 410 nW.

6.4 Measurement Results

The proposed temperature sensor is implemented and fabricated using a standard 180 nm CMOS technology, as shown in Fig. 6.13. The sensing core block has a size of 0.0076 mm^2 and the total circuit including the core, VIC circuit, and 10-bit SAR ADC occupies an area of 0.0768 mm^2. The fabricated chips are tested using a 1.8 V voltage supply in a specially designed thermal chamber and 8 chips are tested in total. The thermal measurement setup

6.4 Measurement Results

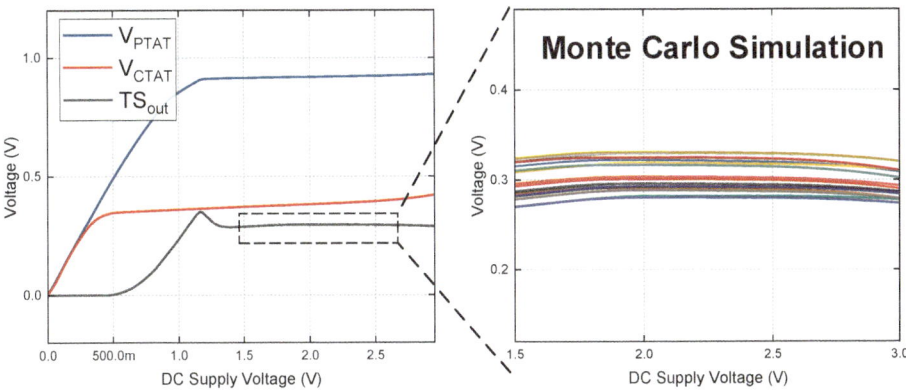

Fig. 6.11 Simulation results of the output voltage of the temperature sensor through voltage supply DC sweep

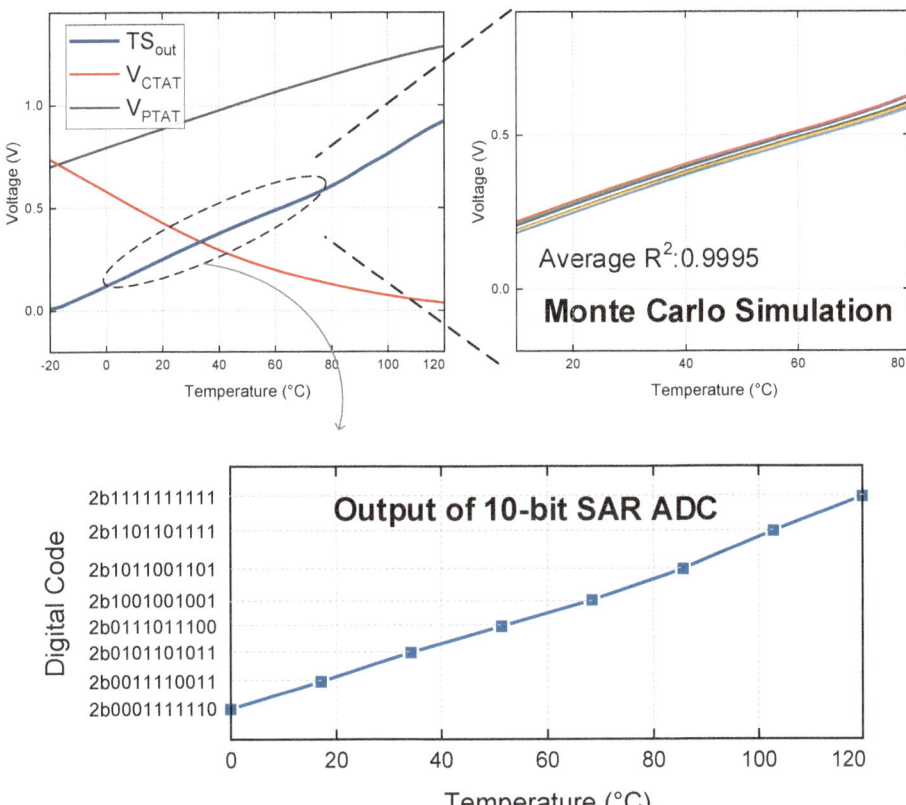

Fig. 6.12 Simulation results of the temperature sensor through temperature sweep

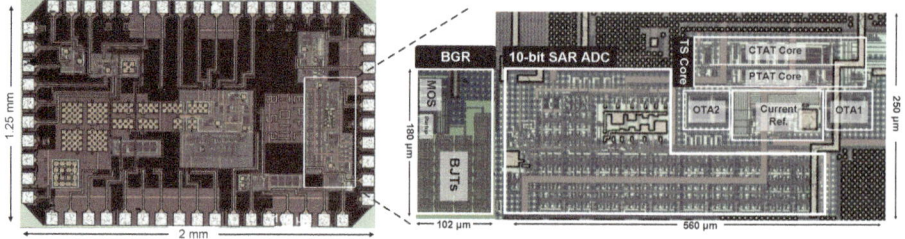

Fig. 6.13 Chip microphotograph of the proposed temperature sensor

comprises two experimental configurations, as shown in Fig. 6.14. The first setup includes a thermo stream featuring an insulated thermal chamber, enabling a temperature sweep spanning from −30 to 110 °C in 1 °C increments. The second setup involves a chamber comprising a thermally isolated flask outfitted with a high-power resistive element, capable of generating the required heat to continuously monitor temperature. The output voltages of the circuit are interfaced with a digital signal acquisition system. In this operational framework, the continuous temperature measurement process entails elevating the temperature of the thermal chamber from room temperature to 110°C by applying a high current to the resistive element. Subsequently, the resistive element is disconnected. As a result, the temperature naturally decreases to room temperature. During this relaxation phase, the output voltage of the proposed sensor is measured. This method can also be applied to measure low temperatures by using a cold spray.

The measured output voltages through the voltage supply DC sweep are shown in Fig. 6.15. It offers a constant output voltage within the valid range of 1.55–2.3 V. The measured results of the temperature sensor output voltage and its related error through

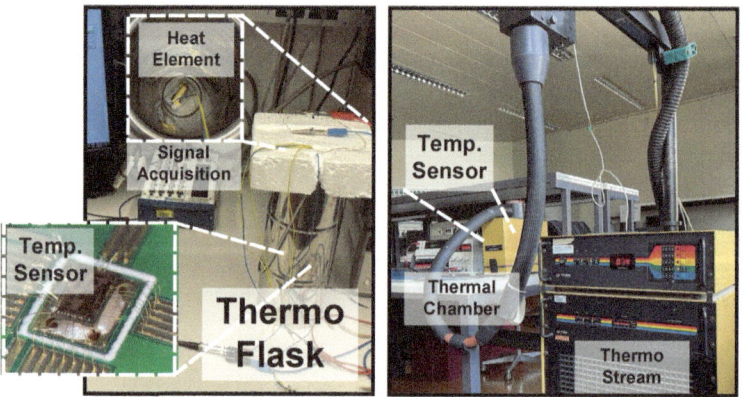

Fig. 6.14 Measurement setup of the presented sensor using a thermal flask (on left) and a thermal chamber with a thermo stream (on right) for temperature sweeping

6.4 Measurement Results

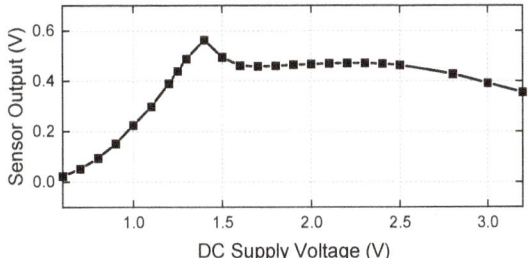

Fig. 6.15 Measurement results of the output voltage of the VIC of the temperature sensor through the voltage supply DC sweep

temperature sweep are shown in Fig. 6.16 with highlighted measurements in the body temperature range for biomedical and wearable applications. The sensor offers a highly linear output over −30 to 110 °C. This characteristic is coupled with high accuracy and sensitivity, as well as a high VTT slope. Employing a two-point calibration technique at −30 and 110 °C, the proposed design achieved an error margin of −0.8/+0.8 °C (worst-case inaccuracy of 1.6 °C) from −30 to 110 °C which presents a 1.1% relative inaccuracy (Rel. IA), as shown in Fig. 6.16.

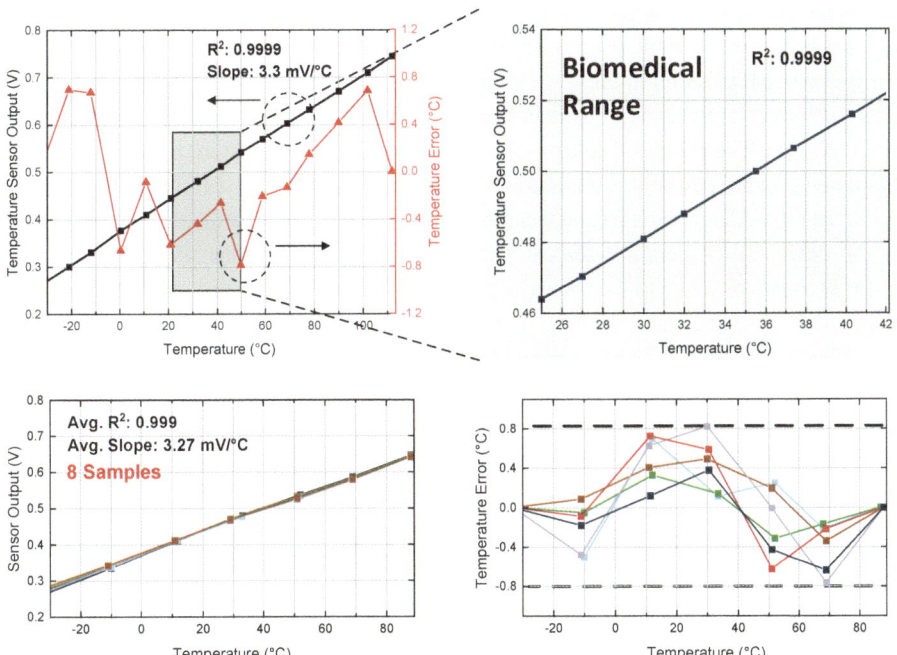

Fig. 6.16 Measured results of multiple samples of temperature sensors and their error through the temperature sweep along with measurements in the body temperature range (brain temperature ranges from 36.5 to 37.5 °C [18])

Table 6.1 Performance comparison with state-of-the-art

Source	CICC [8]	JSSC [20]	JSSC [21]	TCAS-I [13]	SSC-L [19]	This work
Year	2019	2019	2019	2021	2023	
Type	MOS	MOS	MOS	MOS	MOS	MOS
ADC type	FDC	FDC	FDC	FDC	SAR	SAR
Tech. (nm)	65	180	65	130	180	180
Area (mm^2)	0.013	0.044	0.63	0.07	0.0858	0.0768
VDD (V)	0.90	0.80	0.50	0.95	1.20	1.80
Trim points	2	2	2	2	2	2
PP IA (°C)	4.5	2.1	1.6	0.84	4.7	1.6
T_{min} (°C)	−20	−20	0	0	−50	−30
T_{max} (°C)	100	80	100.0	80	130	110
Rel. IA (%)	3.8	2.1	1.6	1.1	2.6	1.1
Power (μW)	0.0006	0.011	0.00011	0.196	0.12	0.432

The measured VTT slope is 3.3 mV/°C. A performance comparison of the proposed design is presented in Table 6.1, considering MOS-based state-of-the-art sensors fabricated in different CMOS technologies. A temperature sensor presented in [19] employs a SAR ADC for conversion purposes, while a frequency-to-digital converter (FDC) is employed in [8, 13, 20, 21]. The proposed sensor exhibits an extensive temperature measurement range with a low relative inaccuracy. Consequently, it offers an effective solution for applications within the biomedical domains and wearable devices. The following factors can be improved for the designed temperature sensor: enhancing the sensor's slope to focus specifically on the biomedical temperature range and offering a larger linear voltage variation due to temperature changes. Implementing a 10-bit SAR ADC inside the chip might consume a large silicon area; alternatively, methods such as voltage-to-time or frequency conversion can be used for greater area efficiency. The performance of the proposed temperature sensor has been compared and summarized in Fig. 6.17 alongside other reported designs from the literature.

6.5 Temperature Sensor for Body Temperature Range

To enhance the resolution and sensitivity of the proposed temperature sensor for the body temperature range, a large VTT slope is required for the digital conversion of small temperature changes by limiting the sensing range from 10 to 45 °C, which aligns with biomedical and body temperature applications. The core design can accordingly follow the circuit model presented in Fig. 6.3, with 7 PTAT cores and 6 CTAT cores. To ensure adequate headroom for

6.6 Summary

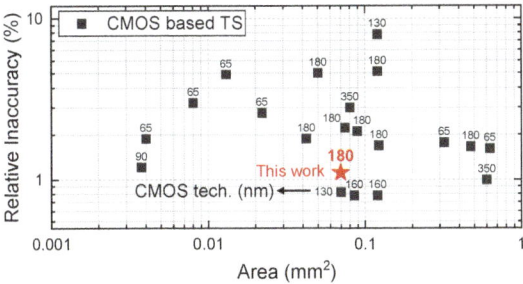

Fig. 6.17 Comparison of the sensor with the reported sensors [22]

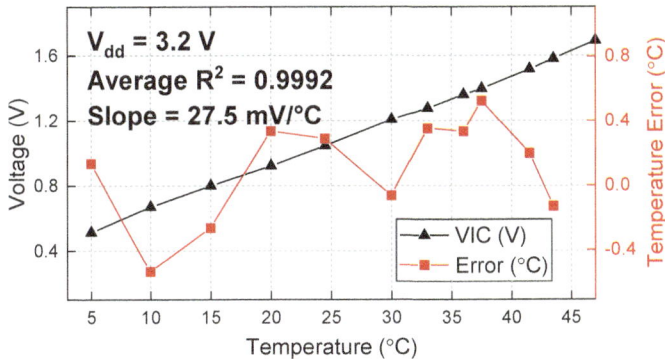

Fig. 6.18 Measured results of the modified temperature sensor and its error through the temperature sweep with a focus on the body temperature range

the circuitry, the sensor should operate on a higher voltage supply. The performance of this biomedical-specific sensor, as depicted in Fig. 6.18, demonstrates a highly linear response, with a slope of 27.5 mV/°C across a temperature range of 5–45 °C. Assuming a reference voltage of 1.8 V, the 10-bit SAR ADC will exhibit a conversion step of 1.7 mV, approximately 15 steps per 1-degree change. Hence, the sensor can precisely detect a temperature increase of 1° within the brain and implants.

6.6 Summary

This chapter presented a compact and energy-efficient CMOS temperature sensor utilizing CTAT/PTAT voltage comparison. The sensor exhibits high linearity across a wide temperature range of −30 to 110 °C and is optimized for low power consumption of 22.3 nW for the sensor core due to its operation in the subthreshold region (432 nW including a 10-bit SAR

ADC). With a small core size of 0.0076 mm² (0.0768 mm² in total) and reliable performance with 2-point calibration and low relative inaccuracy of 1.1%, this sensor is well-suited for integration into small-scale biomedical and wearable devices.

References

1. M. J. Karimi, A. Schmid, and C. Dehollain, "Wireless Power and Data Transmission for Implanted Devices via Inductive Links: A Systematic Review," *IEEE Sensors Journal*, vol. 21, no. 6, pp. 7145–7161, 2021.
2. M. J. Karimi, K. Farhang Razi, C. Dehollain, and A. Schmid, "Modeling and Analysis of a Wirelessly Powered Closed-Loop Implant for Epilepsy," in *2022 IEEE Biomedical Circuits and Systems Conference (BioCAS)*, pp. 414–418, 2022.
3. N. G. Toth, Z. Tang, T. Someya, S. Pan, and K. A. A. Makinwa, "23.7 A BJT-Based Temperature Sensor with ±0.1°C(3σ) Inaccuracy from −55 to 125 °C and a 0.85pJ.K2 Resolution FoM Using Continuous-Time Readout," in *2023 IEEE International Solid- State Circuits Conference (ISSCC)*, pp. 358–360, 2023.
4. B. Park, Y. Ji, and J.-Y. Sim, "A 490-pW SAR Temperature Sensor With a Leakage-Based Bandgap-Vth Reference," *IEEE Transactions on Circuits and Systems II: Express Briefs*, vol. 67, no. 9, pp. 1549–1553, 2020.
5. X. Tang, H. Zhao, and S. Mandal, "A Fully-Integrated 27.12 MHz Inductive Power and Data Telemetry Link for Biomedical Implants," in *2020 IEEE 63rd International Midwest Symposium on Circuits and Systems (MWSCAS)*, pp. 655–658, 2020.
6. Z. Tang, Y. Fang, X.-P. Yu, Z. Shi, L. Lin, and N. N. Tan, "A Dynamic-Biased Resistor-Based CMOS Temperature Sensor With a Duty-Cycle-Modulated Output," *IEEE Transactions on Circuits and Systems II: Express Briefs*, vol. 67, no. 9, pp. 1504–1508, 2020.
7. C.-Y. Lu, S. Ravikumar, A. D. Sali, M. Eberlein, and H.-J. Lee, "An 8b subthreshold hybrid thermal sensor with ±1.07 °C inaccuracy and single-element remote-sensing technique in 22nm FinFET," in *2018 IEEE International Solid - State Circuits Conference—(ISSCC)*, pp. 318–320, 2018.
8. D. S. Truesdell and B. H. Calhoun, "A 640 pW 22 pJ/sample Gate Leakage-Based Digital CMOS Temperature Sensor with 0.25 °C Resolution," in *2019 IEEE Custom Integrated Circuits Conference (CICC)*, pp. 1–4, 2019.
9. P. Chen, Y.-J. Hu, J.-C. Liou, and B.-C. Ren, "A 486k S/s CMOS time-domain smart temperature sensor with −0.85/0.78 °C voltage-calibrated error," in *2015 IEEE International Symposium on Circuits and Systems (ISCAS)*, pp. 2109–2112, 2015.
10. T. Anand, K. A. A. Makinwa, and P. K. Hanumolu, "A VCO Based Highly Digital Temperature Sensor With 0.034 °C/mV Supply Sensitivity," *IEEE Journal of Solid-State Circuits*, vol. 51, no. 11, pp. 2651–2663, 2016.
11. K. Yang, Q. Dong, W. Jung, Y. Zhang, M. Choi, D. Blaauw, and D. Sylvester, "9.2 A 0.6 nJ −0.22/+0.19°C inaccuracy temperature sensor using exponential subthreshold oscillation dependence," in *2017 IEEE International Solid-State Circuits Conference (ISSCC)*, pp. 160–161, 2017.
12. W. Song, J. Lee, N. Cho, and J. Burm, "An Ultralow Power Time-Domain Temperature Sensor With Time-Domain Delta–Sigma TDC," *IEEE Transactions on Circuits and Systems II: Express Briefs*, vol. 64, no. 10, pp. 1117–1121, 2017.
13. J. Li, Y. Lin, N. Ning, and Q. Yu, "A 0.44/−0.4 °C Inaccuracy Temperature Sensor With Multi-Threshold MOSFET-Based Sensing Element and CMOS Thyristor-Based VCO," *IEEE Transactions on Circuits and Systems I: Regular Papers*, vol. 68, no. 3, pp. 1102–1113, 2021.

14. J. Kim, J. Kim, C. Park, M. Yang, and W. Jung, "A –50 to 130 °C, 38.69 pJ/conv Fully Integrated SAR Temperature Sensor Based on Direct Temperature-Voltage Comparison," in *2022 IEEE Asian Solid-State Circuits Conference (A-SSCC)*, pp. 1–3, 2022.
15. B. Razavi, "The Design of a Low-Voltage Bandgap Reference [The Analog Mind]," *IEEE Solid-State Circuits Magazine*, vol. 13, no. 3, pp. 6–16, 2021.
16. B. Razavi, "The StrongARM Latch [A Circuit for All Seasons]," *IEEE Solid-State Circuits Magazine*, vol. 7, no. 2, pp. 12–17, 2015.
17. A. V. Fonseca, P. Maris Ferreira, L. Cron, F. A. P. Barúqui, C. F. T. Soares, and P. Benabes, "A Temperature-Aware Analysis of SAR ADCs for Smart Vehicle Applications," *Journal of Integrated Circuits and Systems*, vol. 13, p. 1–10, Aug. 2018.
18. J. Soukup, A. Zauner, E. M. Doppenberg, M. Menzel, C. Gilman, H. F. Young, and R. Bullock, "The Importance of Brain Temperature in Patients after Severe Head Injury: Relationship to Intracranial Pressure, Cerebral Perfusion Pressure, Cerebral Blood Flow, and Outcome," *Journal of Neurotrauma*, vol. 19, p. 559–571, 2002.
19. J. Kim, J. Kim, C. Park, M. Yang, and W. Jung, "A Wide Range, Energy-Efficient Temperature Sensor Based on Direct Temperature-Voltage Comparison," *IEEE Solid-State Circuits Letters*, vol. 6, pp. 113–116, 2023.
20. T. Someya, A. K. M. M. Islam, T. Sakurai, and M. Takamiya, "An 11-nW CMOS Temperature-to-Digital Converter Utilizing Sub-Threshold Current at Sub-Thermal Drain Voltage," *IEEE Journal of Solid-State Circuits*, vol. 54, no. 3, pp. 613–622, 2019.
21. H. Wang and P. P. Mercier, "A 763 pW 230 pJ/Conversion Fully Integrated CMOS Temperature-to-Digital Converter With 0.81/–0.75 °C Inaccuracy," *IEEE Journal of Solid-State Circuits*, vol. 54, no. 8, pp. 2281–2290, 2019.
22. K. Makinwa, "Smart Temperature Sensor Survey".

System-On-Chip Integration

7

Chapter 4 introduced the circuits designed for remote powering and the power conversion chain of the implanted units, along with their corresponding measurement results. Furthermore, Chap. 5 presented the bi-directional wireless data communication methods, supported by experimental results for each block. To evaluate the overall effectiveness of the system, the proposed circuits were integrated as a system-on-chip (SoC) and fabricated using TSMC 180 nm MM/RF technology. The final version of the system diagram includes all the blocks described in the preceding Chapters. This Chapter begins with a brief introduction and literature review on wirelessly powered SoCs for biomedical applications. The system architecture and the simulation and measurement results are also presented. The characterization of the SoC is achieved using wire bonding and chip-on-board packages. Printed circuit boards (PCBs) have been designed to verify the prototype, as shown in Fig. 7.1.

7.1 Background and Related Research

Certain medical applications require simultaneous data recording and stimulation at multiple body or brain sites while maintaining communication with an external device. To meet this need, a multisite powering and control system is proposed for implant networks, supporting applications such as epilepsy treatment, brain stimulation, cardiac pacing, and spinal cord stimulation. The objective is to facilitate remote power and data transfer across distributed implants. Depending on coverage requirements, multiple-unit and free-floating implants are utilized. Various approaches have been explored for implementing multisite wireless power and data transfer (WPDT). In [1], a dual-frequency heart pacing system was introduced, where two implants operated at distinct carrier frequencies (13.56 and 40.68 MHz) using separate transmitters. However, interference between transmitter coils limits the operating range, and the number of Tx coils must scale with the number of implants, reducing scalability. Another approach in [2] employs physical IDs to individually address implants,

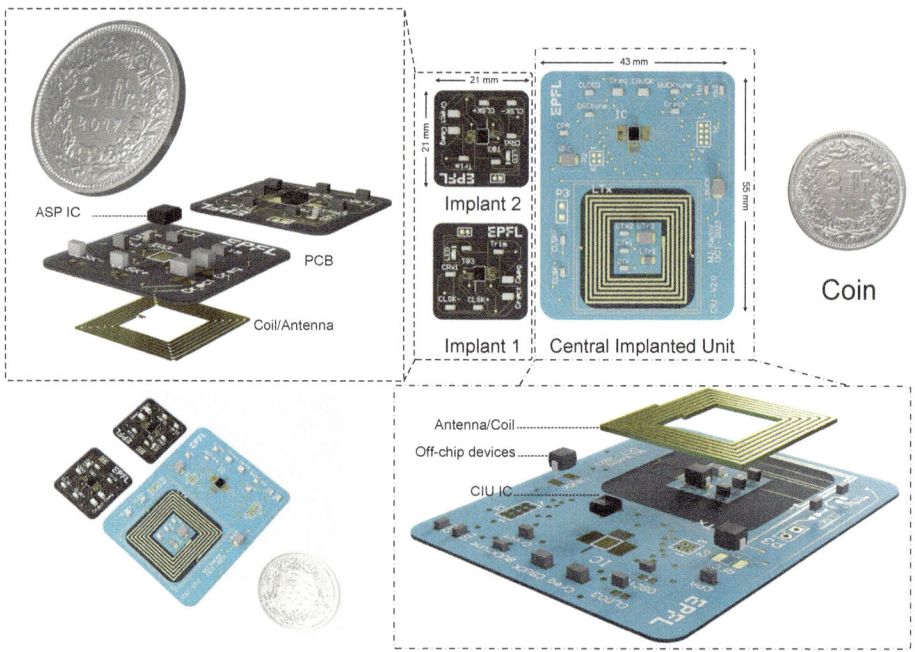

Fig. 7.1 Conceptual view of the implantable boards with the SoCs

enhancing scalability. Additionally, dual-band coil structures have been used to control multiple implants via a single Tx coil at different frequency bands [3], though these systems face challenges in power transfer efficiency and impedance matching. In [4], miniaturized implants were introduced alongside a multisite WPT system that utilizes board-defined passcodes for individual addressing.

To enable independent communication with each implant, multiple access techniques such as time-division multiple access (TDMA) and frequency-division multiple access (FDMA) have been explored. These methods support bidirectional data exchange over a single inductive link while maintaining power transfer efficiency. This work adopts FDMA for both power and data communication with implantable medical devices (IMDs), offering improved efficiency and simplified implementation. By assigning distinct carrier frequencies to each implant, selective communication is achieved, reducing unnecessary power dissipation. Compared to TDMA, which requires complex circuitry and careful idle-state management, FDMA provides a more straightforward solution, allowing precise implant selection while minimizing cross-talk. The FDMA-based WPDT approach dynamically tunes the implant's LC tank resonance through capacitance adjustments. The external device adapts its resonance frequency to match the target implant, ensuring selective power and data transfer. To further enhance PTE, each implant incorporates an automatic resonance tuning system, enabling capacitor fine-tuning for optimized power reception. This selective tuning

7.1 Background and Related Research

mechanism ensures efficient multisite WPDT by delivering power and data exclusively to the intended implant.

Various modulation techniques are employed to enhance the efficiency and reliability of power and data transmission in biomedical applications [5–7]. Among these, amplitude shift keying (ASK) is particularly advantageous for low-power designs, as it eliminates the need for a local oscillator and encodes data by modulating only the amplitude of the carrier signal. Similarly, load shift keying (LSK) is utilized for uplink communication, where the implant modulates the load of the LC tank at the receiver to disrupt the resonant condition of the inductive link. Certain biomedical applications, such as epilepsy treatment, necessitate bi-directional data exchange between an external control unit and the implant. This is achieved through a downlink for transmitting commands and an uplink for receiving recorded neural signals or system parameters. Previous studies have proposed WPDT systems operating at 13.56 or 10 MHz for biomedical applications [8–13], implementing LSK modulation over a single inductive link. For instance [9], introduces a WPDT system incorporating current-modulated energy-reuse back telemetry and an energy-adaptive dual-input low-dropout regulator (LDO), whereas [8] focuses on adaptive power delivery techniques for wireless power transfer (WPT). Additionally, a low-modulation index detectable ASK demodulator tailored for a 13.56 MHz WPDT receiver is presented in [14]. Despite these advancements, research addressing bi-directional multisite WPDT over a single inductive link remains scarce.

To implement the proposed multisite WPDT system, several key challenges must be addressed. First, an efficient WPT mechanism must be developed to reliably and safely power multiple implants deep inside the body or brain. Second, practical constraints, such as variations in distance and alignment between the external unit and implants due to body movements, introduce fluctuations in input voltage and power levels—particularly when utilizing a single Tx unit. Finally, an approach must be established to selectively address each implant while using a shared Tx coil. This Chapter introduces a wirelessly powered, half-duplex bi-directional data communication system tailored for multisite applications, incorporating two implants operating in the 13.56 and 6.78 MHz industrial, scientific, and medical (ISM) bands, as illustrated in Fig. 7.2. To ensure continuous power monitoring, dedicated power control units are implemented for both the central implanted unit (CIU) and autonomous smart patches (ASPs). The system further integrates an automated resonance tuning mechanism [15] to precisely adjust the resonance capacitor, compensating for parasitic capacitances and variations in coupling due to distance and angular misalignment. The downlink communication utilizes ASK modulation, optimizing power consumption in the implanted demodulator, while the uplink employs LSK modulation to induce a slight detuning of the LC tank. This approach maintains stable power transfer with minimal power dissipation in the implants, ensuring uninterrupted wireless power delivery.

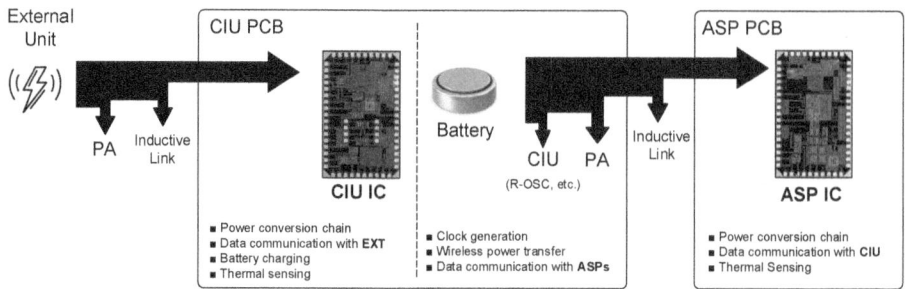

Fig. 7.2 Power flow diagram of the proposed system

7.2 System Architecture

The multisite bi-directional WPDT system is proposed in this Chapter with two important technologies: (1) WPT utilizing dual-band inductive links to supply power to implants positioned at different locations, and (2) a system interfacing with the inductive links to ensure reliable power delivery and bi-directional data transmission for each implant [16]. The block diagram of the proposed system is depicted in Fig. 7.3a. Each implant operates using a dedicated pair of coils, enabling both wireless power and data communication. The system comprises two primary components: (a) the external unit and CIU, which handle wireless power transmission, FSK downlink data, LSK uplink data, and temperature sensing; (b) the CIU and ASPs, responsible for multisite power delivery, ASK downlink data, LSK uplink data, temperature sensing, and automated resonance tuning. In this architecture, the CIU modulates downlink data onto the power waveform using ASK (denoted in red) via the power amplifier (PA) and the generated clock in the oscillator. It also receives uplink data from the implanted unit through LSK modulation (denoted in blue). The bi-directional communication capability, along with the implanted power control unit, enables the external unit/CIU to regulate power input levels dynamically, thereby stabilizing the rectifier output across varying load conditions. Both data transmission and power regulation are managed by an external controller within the implant system. The implanted unit consists of three main functional blocks: (a) power conversion chain (PCC): supplies the necessary voltage and current for the implant's electronic components; (b) data conversion chain (DCC): extracts data and clock signals from the received power waveform and transmits the LSK-modulated signal back to the external unit/CIU; (c) power control unit: ensures that the received power is adequate for system operation and backscattering. The transition between downlink and uplink communication is controlled externally. Initially, power transmission occurs until the implant accumulates sufficient energy for data exchange. Once this threshold is reached, the implant transmits the power state and control bits back to the external unit/CIU. After a predefined number of cycles, command signals are sent as downlink data, freeing the communication link for uplink transmission. The PCC and DCC processes, along with the

7.2 System Architecture

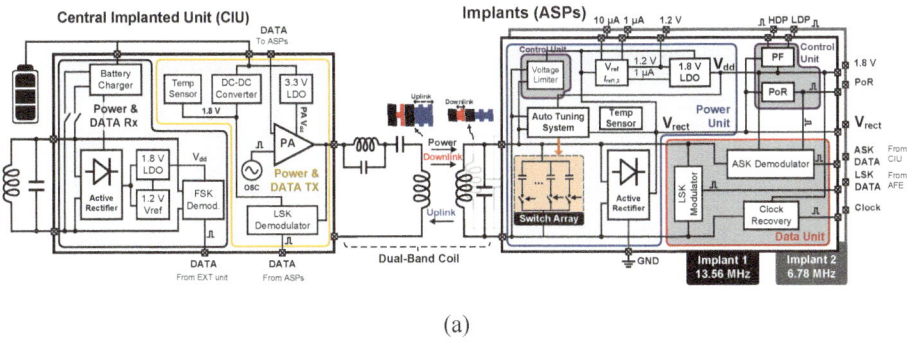

(a)

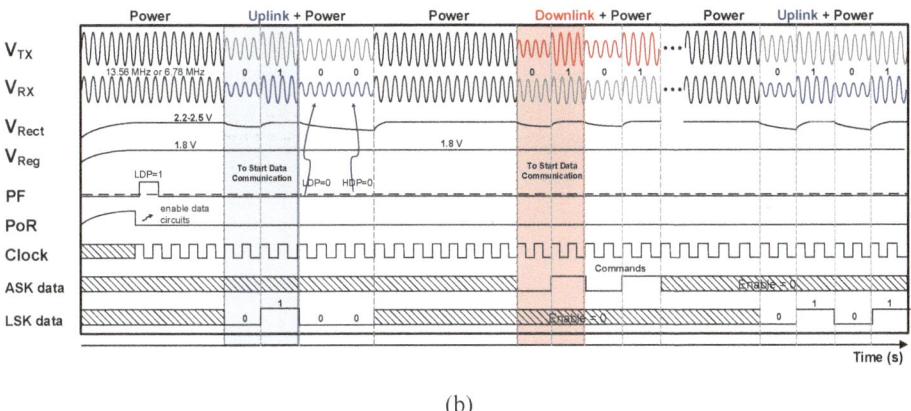

(b)

Fig. 7.3 **a** Block diagram of the proposed wireless telemetry system with a power control unit for closed-loop biomedical implants and **b** timing diagram of the presented system

timing sequence of system operations, are illustrated in Fig. 7.3b. If variations in WPT environmental factors—such as the coupling coefficient or load conditions—lead to excessive or insufficient power delivery, the control system either notifies the external unit/CIU or safeguards the implant by disrupting resonance via a voltage limiter, thereby preventing potential overloading.

The inductive link is designed to operate at two distinct resonance frequencies. The transmitter (Tx) coil is coupled with a capacitor to establish an LC branch. Subsequently, this LC branch is connected in series with an LC tank [3], as depicted in Fig. 7.3. The external unit and CIU supply power to the implant through magnetic coupling, and the induced power must be converted to a reliable DC voltage supply. Designing an efficient and low-power power conversion chain is a challenging task that requires careful consideration of several technical factors, especially for a single inductive link that simultaneously transmits both data and power signals. The overall system-level simulation results for remote powering are shown in Fig. 7.4a. Furthermore, the response of the power control unit, specifically

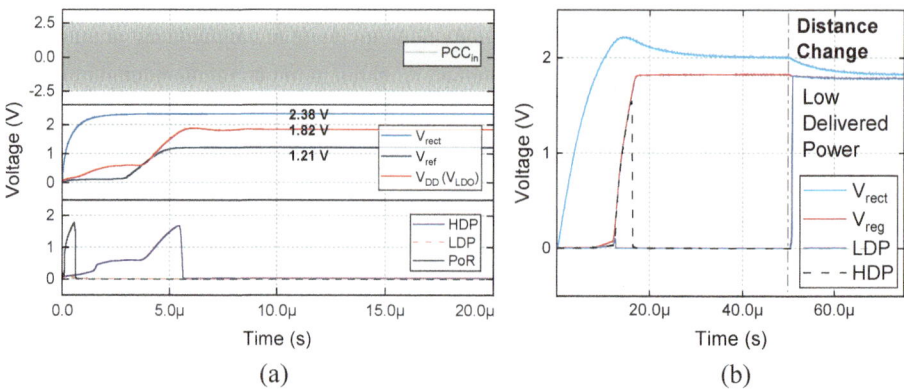

Fig. 7.4 Simulated results of the **a** power conversion chain and **b** the power feedback mechanism, which engages when a reduction in delivered power occurs due to variations in coil distance

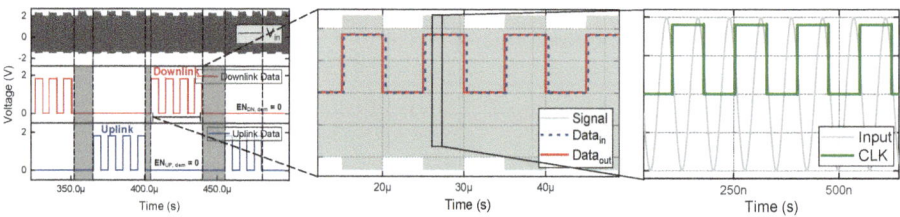

Fig. 7.5 Simulation results of the wireless power transfer and demodulated bi-directional half-duplex downlink and uplink data, along with the recovered clock signal

the activation of the LDP mechanism, to a decrease in delivered power due to variations in coil distance is depicted in Fig. 7.4b. The bi-directional data demodulation process for both downlink and uplink communication, along with clock extraction, is illustrated in Fig. 7.5. During downlink (uplink) data transmission, the LSK (ASK) demodulator at the external unit (implant) must remain inactive, ensuring that $EN_{UP,dem} = 0$ ($EN_{DN,dem} = 0$) to prevent interference and ensure seamless data exchange. Table 7.1 summarizes the overall system performance.

7.3 Measurement Results

The proposed system is designed and fabricated in a 180 nm standard CMOS technology. The CIU and implant chips both have a silicon area of 2.5 mm^2. The die microphotographs are shown in Fig. 7.6. To connect the pads on the chip to the PCB, gold wire bonding with a thickness of 20 μm is utilized with the wedge-wedge type connections. To safeguard against mechanical strain, scratching, and dust, the die and wire bondings are encapsulated

7.3 Measurement Results

Table 7.1 Proposed system-on-chip (SoC) performance summary

CMOS technology		Standard 180 nm
Chip area (CIU/ASP)		2.5/2.5 mm^2
Wireless power transmission		1 CIU + 2 Implants
Coil distance		5–20 mm
Single inductive link: L_{Tx} | L_{Rx}		5.3 µH | 0.42 µH
Wireless power conversion unit		
Carrier frequency		6.78, 13.56 MHz
Rectifier	Input voltage	1.9–3.3 V
	Max VCR and PCE	90.2% & 80.1%
Voltage reference (V_{ref})	Output voltage	1.2
	Voltage variation	0.46 mV/V
	Power dissipation	6.1 µW
Voltage regulator (V_{reg})	Output voltage	1.8 V
	Line regulation	3.07 mV/V
	Load regulation	0.11 mV/mA
	Power dissipation	409 µW
Power control unit		
Power feedback	Control range	2–2.45 V
	Power dissipation	172 µW
Power-on-reset	Threshold	1.75 V
	Power dissipation	72 µW
Voltage limiter	Threshold	3.53 V
	Static power	0.7 µW
Automatic resonance tuning system		
Total on-chip Cap.		75 pF
Resolution and tuning bits		5 pF | 6 CNTRL bits
Tuning range		±15%
Tuning method		Monotonic sweep
Power consumption		154.7 µW
Data communication unit		
Modulation (downlink | uplink)		FSK|ASK & LSK
Power consumption		5.06 µW

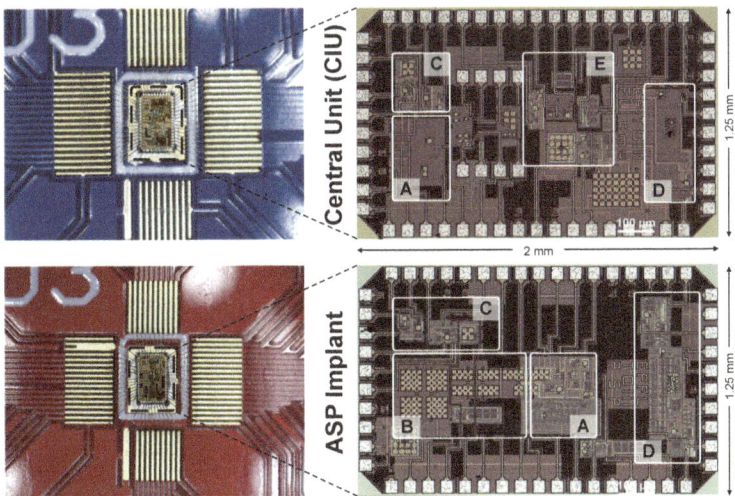

Fig. 7.6 Chip microphotograph of the proposed CIU and ASP SoCs, **a** power conversion chain, **b** power control unit, **c** data communication unit, **d** temperature sensing unit, **e** wireless power transfer unit

with a non-conductive glob top. The glob top serves the purpose of providing electrical insulation, mechanical support, and a controlled environment that protects against light and moisture, thereby facilitating better characterization. Two configurations of the inductive link have been implemented in the system: one with implant coils mounted on separate boards and another with both coils integrated onto a single board. The former configuration allows placement on either side of the external unit's coil, providing flexibility in system deployment. The received voltage at the chip is regulated to a maximum of 3.3 V, while its

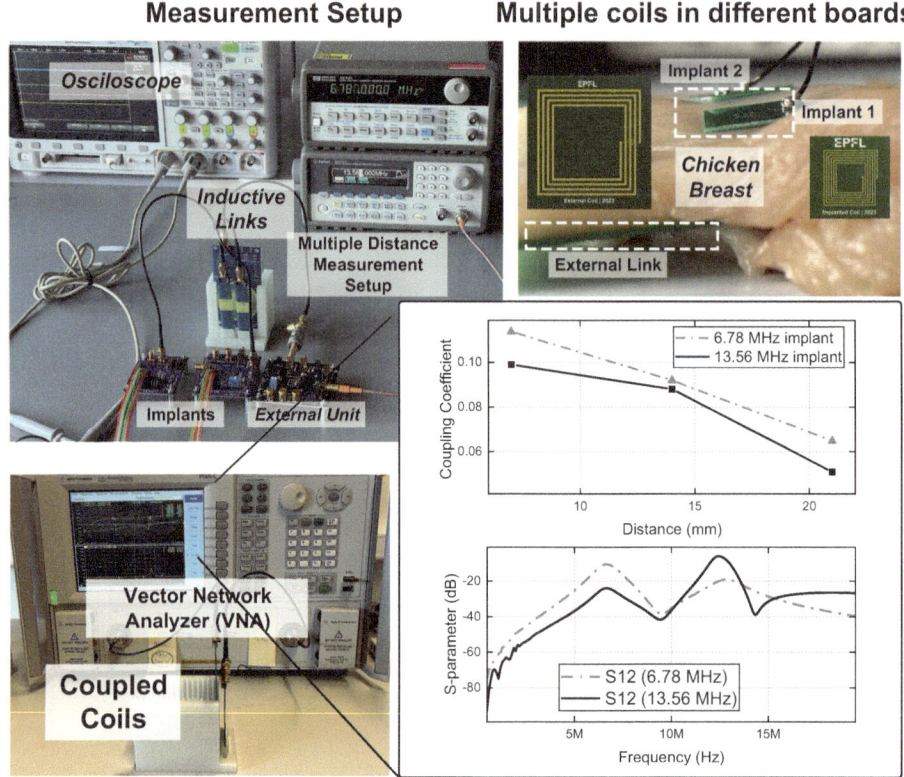

Fig. 7.7 Measurement setup for the SoC and inductive links in various distance values

LDOs supply a stable 1.8 V output. To validate the prototype, PCBs have been designed, as depicted in Fig. 7.7. The experimental setup includes an oscilloscope and a multimeter for measurement and verification. Additionally, external capacitors of 40 and 800 nF are required for the rectifier and the 1.8 V LDO, respectively. The inductive link consists of a Tx coil with an inductance of 5.3 µH and an outer diameter of 43 mm, along with implant Rx coils featuring an inductance of 420 nH and an outer diameter of 13.5 mm. These coils are positioned with a separation distance of 5–20 mm in air and 1 cm within biological tissue (chicken breast), as illustrated in Fig. 7.7 in the multiple-distance measurement setup.

The rectifier's PCE measurement is conducted by placing a 10 Ω resistor in series with the rectifier and measuring the input power using two differential active probes. The rectifier achieves a maximum VCR of 90.2% and a peak PCE of 80.1%. Figure 7.8 presents the measurement results of the proposed PCC unit in the absence of data communication. The transient performance of the ART system, following WPT initiation and resonance frequency optimization, is shown in Fig. 7.9, with a tuning time of 448 µs.

7.3 Measurement Results

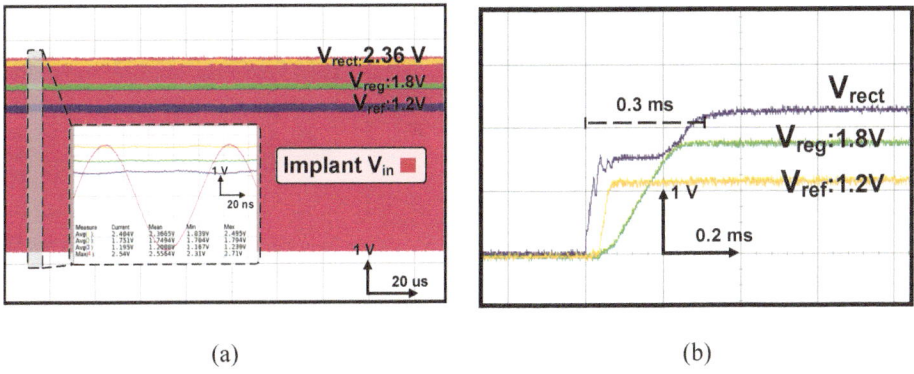

Fig. 7.8 Measured waveform of the **a** power conversion chain (PCC) including V_{rect}, V_{reg}, and V_{ref}, **b** transient and startup of the PCC

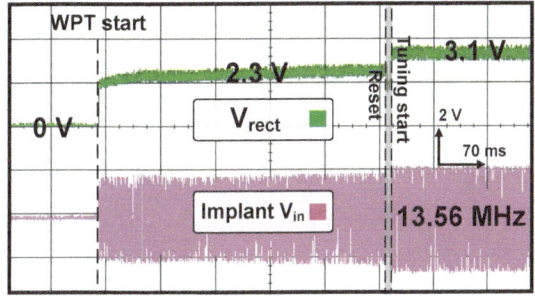

Fig. 7.9 Measurement results of transient received input sine voltage and the rectified voltage after starting WPT and automatic tuning

A pulse modulated using ASK is continuously transmitted to the DCC unit via the inductive link. To assess the ASK demodulator's performance in the ASPs, a 100-bit pseudorandom data stream is generated at a 100 Hz sampling rate. The ASK-modulated signal is then received at the DCC unit as downlink data. The demodulator processes and converts data within a minimum of 20 cycles (10 cycles in simulation), limiting the maximum data rate to $f_c/10$, where f_c denotes the carrier frequency (e.g., 13.56 or 6.78 MHz). Figure 7.10a illustrates simultaneous wireless power transfer and downlink data communication, along with clock extraction in the WPDT system. These signals are inductively transmitted from the central unit to an implant. The received implant voltage, as well as the corresponding DCC and PCC unit inputs, are depicted, highlighting the rectified, regulated, and reference voltage levels.

Figure 7.10b presents the LSK uplink data transmission from a patch implant to the central unit. A minimum rectifier voltage of 1.9 V must be maintained to ensure stable operation. To facilitate multisite power and data transmission across multiple implants, the oscillator in the CIU must be tuned to two distinct frequencies, aligning its resonance frequency with that of the implants. Initially, a 13.56 MHz signal is generated by the primary unit to activate implant 1. Wireless power and bi-directional data transmission then proceed through the

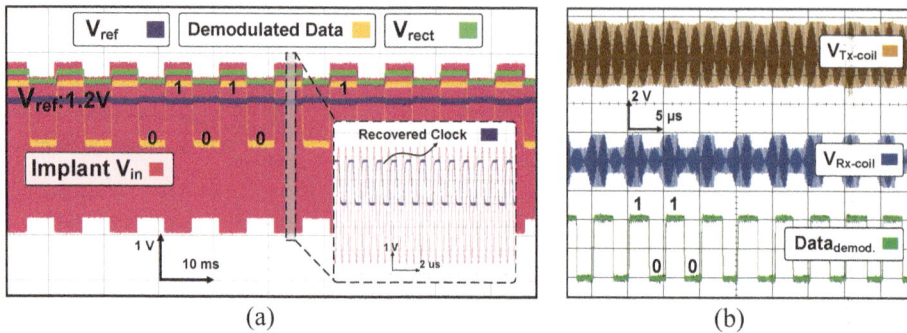

Fig. 7.10 Experimental results of the **a** wireless power and downlink data transmission at 13.56 MHz, **b** LSK data demodulation as uplink communication

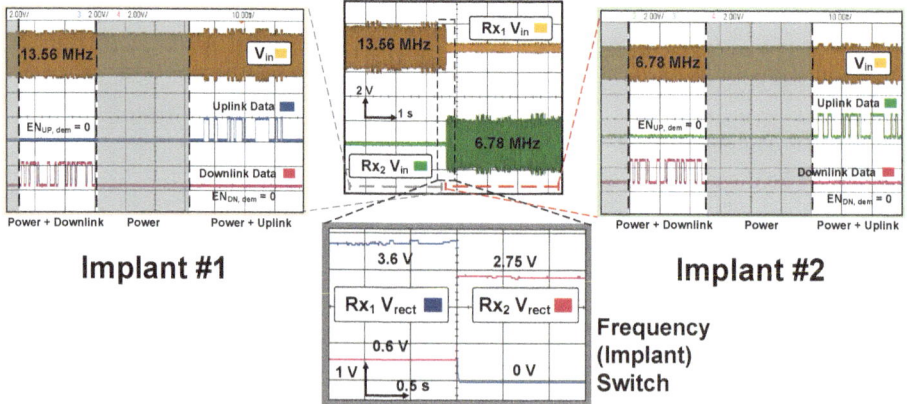

Fig. 7.11 Measurement results of the multi-site wireless power and bi-directional data transmission

inductive link until the frequency of the primary unit is switched to 6.78 MHz, activating implant 2 and enabling WPDT. Figure 7.11 presents the measured results of the multisite WPDT system, enabling both downlink and uplink data communication.

7.4 Discussion

The important features and parameters of the proposed WPDT system are compared to the state-of-the-art wireless implants in Tables 7.2 and 7.3. They include their power and data conversion performance along with the power consumption and efficiency. The proposed system has the advantage of bi-directional data transmission, compared with conventional one-directional systems. It also offers an enhanced efficiency of the WPDT system in comparison to existing bi-directional WPDT systems [17, 18]. The proposed system also pro-

7.4 Discussion

Table 7.2 Comparison of recently published WPDT systems

Reference	TBCAS'16 [19]	JSSC'16 [7]	TBCAS'17 [17]	JSSC'18 [20]	JSSC'18 [18]	JSSC'19 [6]	JSSC'20 [14]	TBCAS'21 [5]	TMTT'22 [4]	TBCAS'22 [8]	JSSC'23 [9]	This work
Technology (nm)	180	65	600	65	180	180	65	180 BCD	180	180	180	180
WPT topology	Passive Rect.+ LDO	Active Rect.	Passive Rect. + LDO	Active Rect.	Active Multiplier + LDO	Passive + Active Rect.	Active Rect. + DC-DC	Passive Rect. + LDO	Passive Rect. + LDO	Active Rect.	Active Rect. + LDO	Active Rect. + LDO
f_{power} (MHz)	10	13.56	13.56	13.56	13.56	1.7–30	13.56	6.5	40.68	13.56	13.56	13.56 & 6.78
Data Rate$_{DN}$(Mbps)	–	–	0.4	–	0.211	–	0.1	2.5	0.01	–	–	0.1
DN-modulation	–	–	OOK	–	BPSK	–	ASK	FSK	ASK-PWM	–	–	ASK
Data Rate$_{UP}$(Mbps)	2	6.78	1.35	–	0.105	5	–	–	–	0.212	1	0.678
Back telemetry	LSK	COOK[†]	PPSK+	LSK	LSK	L-RSK	–	–	–	LSK	LSK	LSK
Distance (mm)	9	10	5–15	6	10	–	–	5	80	6	10	5–20
Efficiency (%)	–	–	60.64	70.6	–	80	75.4	89.6	–	65.6	75	66
V_{in} Peak (V)	–	1.7	–	–	–	–	–	–	3.8	–	2.45	3.3
Max. P_{out} (mW)	15	–	8*	49.4	7.43*	93.8	9.2	115	–	162	2	36
Area (mm^2)	0.28	0.92	–	1.44/1.44	1.6/25	1.87/1.37	5.11	2.32	1.2	2.22/2.16	2.52	0.82
Rectifier V_{out} (V)	1.15–2	1.2	–	–	2	–	–	–	–	–	2	2.2–3.3
Max. PCE_{Rect} (%)	–	–	–	–	80.8	–	75.4	–	–	–	83.3	80.1
VDD (V)	0.925	–	5	1.2–2.5	1.8	–	1/1.5	–	1.8 V	3.3	1.8	1.8
Control unit	–	–	Power regulation + PoR	Tx/Rx regulation	–	Controlled Tx	Current sensor	–	VL, PoR	Rect. Rx + Mode-Ctrl Tx	Adaptive LDO	ART + PF + VL + PoR

DN: downlink, UP: uplink, † COOK: cycling on-off shift keying, + PPSK: passive phase shift keying

Table 7.3 Comparative benchmarking of state-of-the-art WPDT systems [16]

	This work	TMTT'22 [4]	JSSC'22 [2]	Sci. Report'20 [1]	JSSC'18 [18]	TBCAS'17 [17]
Target application	Neural	Neural/cardiac	Neural/cardiac	Cardiac	Neural	–
CMOS (nm)	180	180	180	180	180	600
Power source	Resonant inductive	Resonant inductive	Magnetoelectric	Resonant inductive	Resonant inductive	Resonant inductive
Carrier frequency (MHz)	13.56, 6.78	40.8	0.33	13.56, 40.68	13.56	13.56
Multisite strategy	1 TX + 2 implants	1 TX + 2 implants	1 TX + 2 implants	2 TXs + 2 implants	1 Tx + 1 implant	1 Tx + 1 implant
Individual addressability	Different f_C	PCB Passcode	PUF ID	Different f_C	–	–
SoC power (mW)	1.76	27 µW	9 µW	3.3 µW	3.12	–
Chip area (mm²)	2×2.5	0.75×1.6	1×0.8	0.85×0.45	5×5	21.42
Tx-implant distance (mm)	5–20	5–80	5–40	5–60	10	5–15
Link geometry	Planar PCB coil 43/13.5 mm 5.3/0.42 µH	Planar PCB coil 35/14 mm 1.96/1.91 µH	Planar PCB coil 2×3 mm²	Planar PCB coil N.A./11 mm 1.8 µH	High-Q ferrite coil	Copper wire 25 mm/16 mm 0.21/1.55 µH
Off-chip components	coil + 3	coil + 3	ME film + 1	coil + 3	coil + caps (≥5)	N.A.
Data telemetry	Downlink: ASK Uplink: LSK 0.1/0.678 Mbps	Only downlink: ASK-PWM 10 kbps	Only downlink: ASK 5.16 kbps	Only downlink: ASK	Downlink: BPSK Uplink: LSK 211/105 kbps	Downlink: OOK Uplink: PPSK 0.4/1.35 Mbps
Max. Pout (mW)	36	–	–	–	7.43	100
Rectifier	Active rectifier Max. VCR: 90.1 % Max. PCE: 80.1 %	Passive rectifier	Active rectifier	N.A.	Active rectifier Max. PCE: 80.8%	Half-wave rectifier
LDO (VDD)	1.8 V 3.07 mV/V	1.8 V 18 mV/V	1 V	N.A.	1.8 V	5 V
Control unit	Power feedback 2–2.45 V PoR Voltage limiter ≥3.53 V	PoR voltage limiter	PoR	No	No	Power regulation PoR
Adaptive unit	Automatic tuning Monotonic sweeping 6 CNTR bits, 75 pF coverage	No	Adaptive power converter	No	No	No

7.4 Discussion

Table 7.4 Characterisations of the presented circuits

Design block	Parameters	Value	Design block	Parameters	Value
Rectifier	Max. VCR	93%	PoR	Power (OFF/ON)	0.073/0.269 mW
	Max. PCE	80.8%		Silicon area	0.000143 mm^2
	Off-chip capacitor	2.5 nF		Voltage threshold	2.2 V
	Silicon area	0.036 mm^2	Power feedback	Power consumption	0.172 mW
	Input voltage	1.5–4 V		Silicon area	0.0173 mm^2
Voltage refernece	Output voltage	1.2 V		Voltage threshold	2 and 2.45 V
	Power consumption	0.053 mW	Voltage limiter	Power consumption	0.0007 mW
	Line variation	0.466 mV/V		Silicon area	0.00138 mm^2
	Silicon area	0.00775 mm^2		Voltage threshold	3.53 V
LDO regulator	Input voltage	2–4 V	DC-DC converter	Output voltage	1.8 V
	Quiescent current	3 uA		Power efficiency	92%
	PSR	−63 dB at 100 Hz		Off-chip components	2
	Output voltage	1.8 V		Silicon area	0.026 mm^2
	Quiescent power	0.098 mW	ASK Demodulator	Max. data rate	2.7 Mbps
	Load regulation	0.03 mV/mA		Min. modulation index	10%
	Line regulation	0.485 mV/V		Power consumption	0.045 mW
	Silicon area	0.07 mm^2		Silicon area	0.00144 mm^2
Auto tuning system	Power consumption	0.155 mW	FSK demodulator	Max. Data rate	1.3 Mbps
	Total on-chip capacitors	75 pF		Min. modulation index	10%
	Tuning clock	53 kHz		Power consumption	0.69 mW
	Tuning bits and range	6 control bits, ±15%		Silicon area	0.015 mm^2
	Tuning method	Monotonic sweep	Temperature sensor	Relative inaccuracy	1.1%
	Resolution	5 pF		Temperature range	−30 to 110
	Silicon area	0.339 mm^2		V-T slope	3.3 mV/C
Relaxation oscillator	Power consumption	0.39 mW		Power consumption	23 nW
	Silicon area	0.0162 mm^2		Data conversion	10-bit SAR ADC
	Frequency range	6.78–13.56 MHz		Silicon area	0.0247 mm^2
Power amplifier	Topology	Class-E PA			
	Off-chip components	2			
	Silicon area	0.00625 mm^2			

vides wireless powering to multisite implants. Further improvements within the system can address the following factors effectively. Enhancing power efficiency can be achieved by combining rectification and voltage regulation as a regulating rectifier that offers higher overall efficiency. Moreover, the presented power feedback unit requires an additional controller to adjust the power. However, this process could be handled by implementing an autonomous analog circuit directly connected to the power amplifier, which is capable of handling the power adjustments autonomously. The implants proposed in [2] operate on the principle of physically unclonable functions (PUF), which requires additional native layers for fabrication, and complex circuitry to ensure process, voltage and temperature (PVT) stability. The implants in [1] have a drawback related to the increased count of Tx coils, which leads to interference issues among these coils.

The summary of the characterization of the proposed blocks and circuits is presented in Table 7.4, including all wireless power, data communication, and thermal sensing circuits and systems. Within the system's operation, not every designed block needs to remain powered ON at all times. Depending on the particular operational phase, each unit is activated as needed for its function. Among these circuits, the power conversion chain remains continuously powered, constituting the most power-consuming components, including the rectifier and the voltage regulator for the ASPs and wireless power transfer unit in the CIU. Additional control circuits, like the tuning system and data communication units, are activated as per necessity. These units consume a smaller portion of the overall power consumption compared to the continuously powered components like the power conversion chain.

7.5 Summary

This Chapter introduces a wirelessly powered bi-directional data communication system tailored for multisite implanted biomedical applications. The system enables independent wireless power delivery and data exchange for each implant through inductive resonant links utilizing dual-band coils. It ensures efficient power transfer while incorporating digitally-assisted active rectification, power control feedback, and an automated resonance tuning mechanism with six control bits and 75 pF capacitance tuning. A key contribution of this work is the integration of a self-sampling separated-V_b ASK demodulator, an averaging LSK demodulator, an FSK demodulator, and a clock recovery circuit to facilitate data conversion in both the external and implanted units. Experimental validation using multisite inductive links demonstrates successful wireless power transmission and bi-directional data communication, achieving a maximum overall power transfer efficiency of 31.2% and a peak data rate of 1.1 Mbps.

References

1. H. Lyu, M. John, D. Burkland, B. Greet, A. Post, A. Babakhani, and M. Razavi, "Synchronized Biventricular Heart Pacing in a Closed-chest Porcine Model based on Wirelessly Powered Leadless Pacemakers (Scientific Reports, (2020), 10, 1, (2067), 10.1038/s41598-020-59017-z)," *Scientific Reports*, vol. 10, no. 1, pp. 1–9, 2020.
2. Z. Yu, J. C. Chen, Y. He, F. T. Alrashdan, B. W. Avants, A. Singer, J. T. Robinson, and K. Yang, "Magnetoelectric Bio-Implants Powered and Programmed by a Single Transmitter for Coordinated Multisite Stimulation," *IEEE Journal of Solid-State Circuits*, vol. 57, no. 3, pp. 818–830, 2022.
3. M.-L. Kung and K.-H. Lin, "Enhanced Analysis and Design Method of Dual-Band Coil Module for Near-Field Wireless Power Transfer Systems," *Microwave Theory and Techniques, IEEE Transactions on*, vol. 63, pp. 821–832, 03 2015.
4. I. Habibagahi, J. Jang, and A. Babakhani, "Miniaturized Wirelessly Powered and Controlled Implants for Vagus Nerve Stimulation," *Digest of Papers—IEEE Radio Frequency Integrated Circuits Symposium*, vol. 2022-June, pp. 51–54, 2022.
5. Y. Park, S. T. Koh, J. Lee, H. Kim, J. Choi, S. Ha, C. Kim, and M. Je, "A Wireless Power and Data Transfer IC for Neural Prostheses Using a Single Inductive Link with Frequency-Splitting Characteristic," *IEEE Transactions on Biomedical Circuits and Systems*, vol. 15, no. 6, pp. 1306–1319, 2021.
6. J. Pan, A. A. Abidi, W. Jiang, and D. Marković, "Simultaneous Transmission of Up To 94-mW Self-Regulated Wireless Power and Up To 5-Mb/s Reverse Data Over a Single Pair of Coils," *IEEE Journal of Solid-State Circuits*, vol. 54, no. 4, pp. 1003–1016, 2019.
7. S. Ha, C. Kim, J. Park, S. Joshi, and G. Cauwenberghs, "Energy Recycling Telemetry IC with Simultaneous 11.5 mW Power and 6.78 Mb/s Backward Data Delivery over a Single 13.56 MHz Inductive Link," *IEEE Journal of Solid-State Circuits*, vol. 51, no. 11, pp. 2664–2678, 2016.
8. G. Namgoong, W. Park, and F. Bien, "A 13.56 MHz Wireless Power Transfer System With Fully Integrated PLL-based Frequency-Regulated Reconfigurable Duty Control for Implantable Medical Devices," *IEEE Transactions on Biomedical Circuits and Systems*, vol. 16, no. 6, pp. 1116–1128, 2022.
9. M. Kim, H. S. Lee, J. Ahn, and H. M. Lee, "A 13.56-MHz Wireless Power and Data Transfer System With Current-Modulated Energy-Reuse Back Telemetry and Energy-Adaptive Voltage Regulation," *IEEE Journal of Solid-State Circuits*, vol. 58, no. 2, pp. 1–11, 2022.
10. B. Lee, M. Kiani, and M. Ghovanloo, "A Triple-Loop Inductive Power Transmission System for Biomedical Applications," *IEEE Transactions on Biomedical Circuits and Systems*, vol. 10, pp. 138–148, 2016.
11. M. J. Karimi, S. Mehdi, C. Dehollain, and A. Schmid, "Wireless Power and Data Transceiver in A Central Implanted Unit for Biomedical Applications," in *2024 IEEE 15th Latin America Symposium on Circuits and Systems (LASCAS)*, pp. 1–5, 2024.
12. Y.-P. Lin, C.-Y. Yeh, P.-Y. Huang, Z.-Y. Wang, H.-H. Cheng, Y.-T. Li, C.-F. Chuang, P.-C. Huang, K.-T. Tang, H.-P. Ma, Y.-C. Chang, S.-R. Yeh, and H. Chen, "A Battery-Less, Implantable Neuro-Electronic Interface for Studying the Mechanisms of Deep Brain Stimulation in Rat Models," *IEEE Transactions on Biomedical Circuits and Systems*, vol. 10, no. 1, pp. 98–112, 2016.
13. J. Zhao, L. Yao, R.-F. Xue, P. Li, M. Je, and Y. P. Xu, "An Integrated Wireless Power Management and Data Telemetry IC for High-Compliance-Voltage Electrical Stimulation Applications," *IEEE Transactions on Biomedical Circuits and Systems*, vol. 10, no. 1, pp. 113–124, 2016.
14. D. Ye, Y. Wang, Y. Xiang, L. Lyu, H. Min, and C. J. Shi, "A Wireless Power and Data Transfer Receiver Achieving 75.4% Effective Power Conversion Efficiency and Supporting 0.1% Modulation Depth for ASK Demodulation," *IEEE Journal of Solid-State Circuits*, vol. 55, no. 5, pp. 1386–1400, 2020.

15. M. J. Karimi, M. Jin, C. Dehollain, and A. Schmid, "A Wireless Power Conversion Chain With Fully On-Chip Automatic Resonance Tuning System for Biomedical Implants," *IEEE Open Journal of Circuits and Systems*, vol. 5, pp. 117–127, 2024.
16. M. J. Karimi, M. Jin, Y. Zhou, C. Dehollain, and A. Schmid, "Wirelessly Powered and Bi-directional Data Communication System with Adaptive Conversion Chain for Multisite Biomedical Implants Over Single Inductive Link," *IEEE Transactions on Biomedical Circuits and Systems*, pp. 1–11, 2024.
17. D. Jiang, D. Cirmirakis, M. Schormans, T. A. Perkins, N. Donaldson, and A. Demosthenous, "An Integrated Passive Phase-Shift Keying Modulator for Biomedical Implants With Power Telemetry Over a Single Inductive Link," *IEEE Transactions on Biomedical Circuits and Systems*, vol. 11, no. 1, pp. 64–77, 2017.
18. C.-H. Cheng, P.-Y. Tsai, T.-Y. Yang, W.-H. Cheng, T.-Y. Yen, Z. Luo, X.-H. Qian, Z.-X. Chen, T.-H. Lin, W.-H. Chen, W.-M. Chen, S.-F. Liang, F.-Z. Shaw, C.-S. Chang, Y.-L. Hsin, C.-Y. Lee, M.-D. Ker, and C.-Y. Wu, "A Fully Integrated 16-Channel Closed-Loop Neural-Prosthetic CMOS SoC With Wireless Power and Bidirectional Data Telemetry for Real-Time Efficient Human Epileptic Seizure Control," *IEEE Journal of Solid-State Circuits*, vol. 53, no. 11, pp. 3314–3326, 2018.
19. Y. P. Lin and K. T. Tang, "An Inductive Power and Data Telemetry Subsystem with Fast Transient Low Dropout Regulator for Biomedical Implants," *IEEE Transactions on Biomedical Circuits and Systems*, vol. 10, pp. 435–444, 2016.
20. C. Huang, T. Kawajiri, and H. Ishikuro, "A 13.56-MHz Wireless Power Transfer System With Enhanced Load-Transient Response and Efficiency by Fully Integrated Wireless Constant-Idle-Time Control for Biomedical Implants," *IEEE Journal of Solid-State Circuits*, vol. 53, no. 2, pp. 538–551, 2018.

Summary and Conclusions 8

In summary, this Book has developed an innovative multisite implantable wireless system for neural recording and stimulation, consisting of a three-layer architecture for patient autonomy. The system includes a wireless power and data platform using an inductive link, enabling continuous and simultaneous power delivery and bidirectional data communication. The external unit, implemented into a headstage, facilitates communication with the internal base station, located in a burr hole, which wirelessly delivers power and data to smart patches implanted on the cortex surface. Each smart patch, equipped with electrodes and signal conditioning electronics, records cortical activity, detects seizures, and electrically stimulates the cortex for intervention.

In Chap. 2, a system-level model has been proposed for a WPDT system with a multi-channel iEEG signal recording, data compression, and a stimulation unit. The WPDT units are analyzed and modeled to a generic system-level block.

In Chap. 3, the design of a dual-band inductive link for wireless power and data transmission was presented. It focuses on how the shape of coils affects dual-band wireless power transmission. Circular, square, and octagonal spiral coils were analyzed.

In Chap. 4, the design and characterization of a wireless power conversion chain and transmission platform were presented. Various techniques for circuit and system design were presented that enable efficient remote powering using a single inductive link. Also, the key components for energy harvesting are presented, including the digitally-assisted active rectifier, voltage reference, voltage regulator, and power control units, as well as the class-E power amplifier and the relaxation oscillator for wireless power transfer.

In Chap. 5, the design and characterization of the data communication units were presented. Different modulation techniques were proposed that achieve low-power data and clock recovery using a single inductive link in the CIU and ASPs.

In Chap. 6, a compact and energy-efficient CMOS temperature sensor was designed. It utilizes a CTAT/PTAT voltage comparison. The sensor exhibits high linearity across a wide temperature range of −30 to 110 °C. It is optimized for a low power consumption of 22.3 nW due to its operation in the subthreshold region. It has a small core size of 0.0076 mm^2 and reliable performance with 2-point calibration and low relative inaccuracy of 1.1%.

In Chap. 7, a wirelessly powered and bi-directional data communication system as a system-on-chip designed for multisite IMDs. It supports simultaneous wireless power and data communication along with measurement results for each implant using inductive links with dual-band coils.

GPSR Compliance

The European Union's (EU) General Product Safety Regulation (GPSR) is a set of rules that requires consumer products to be safe and our obligations to ensure this.

If you have any concerns about our products, you can contact us on ProductSafety@springernature.com

In case Publisher is established outside the EU, the EU authorized representative is:

Springer Nature Customer Service Center GmbH
Europaplatz 3
69115 Heidelberg, Germany

Batch number: 08714815

Printed by Printforce, the Netherlands